The Language of Genes

Steve Jones

The Language of Genes

Solving the
Mysteries of Our
Genetic Past,
Present and Future

ANCHOR BOOKS
A DIVISION OF
RANDOM HOUSE, INC.
New York

FIRST ANCHOR BOOKS TRADE PAPERBACK EDITION, JULY 1995

Library of Congress Cataloging-in-Publication Data

Jones, Steve, 1944–
The language of genes: solving the mysteries of our genetic past, present and future
Steve Jones.—1st Anchor Books ed.
p.   cm.
Includes bibliographical references and index.
1. Human genetics—Popular works.   I. Title.
QH431.J575   1994
573.2'1—dc20   93-44626
CIP

Anchor ISBN: 978-0-385-47428-3

www.anchorbooks.com

Printed in the United States of America
10   9   8   7

To my parents and my brother,
who share my genes and my affections

Contents

PREFACE

A Malacologist's Apology

I HAVE SPENT—some might say wasted—most of my scientific career working on snails. A malacologist might seem a surprising author for a book about human genetics. However, my research, when I was still able to do it, was not driven by a deep interest in mollusks. Indeed, one of the few occasions when I thought of giving up biology as a career was when I first had to dissect one. Twenty-five years ago snails were among the few creatures whose genes could be used to study evolution. They carry a statement of ancestry on their shells in the form of inherited patterns of color and banding. By counting genes in different places and trying to relate them to the environment one could get an idea of how and why snail populations diverged from each other: of how and why they evolved.

At the time, the idea that it might ever be possible to do this with humans seemed absurd. Genetics textbooks of the 1960s were routine things. They dealt with the inheritance of pea shape, the sex lives of fungi and the new discoveries about the molecular biology of viruses and their bacterial hosts. Of humans, there was scarcely a mention—usually just a short chapter tagged on at the end which described pedigrees of abnormalities such as hemophilia or color blindness.

Part of this reticence about human genetics was due to ignorance but part came from the dismal history of the subject. In its early days, the study of human inheritance was the haunt of charlatans, most of whom had a political ax to grind. Absurd pedigrees claiming to show the inheritance of criminality or of genius were the norm. Ignorance and confidence went together. Many biologists promoted the idea that it was possible to improve the human race by selective breeding or by the elimination of the genetically unfit. The adulteration of human genetics reached its disastrous end in the Nazi experiment, and for many years it was seen as at best unfashionable to discuss the nature of inherited differences among people.

After the Second World War, the United Nations was involved in publishing a book—*Statement on Race*, by the American anthropologist Ashley Montagu—which tried to kill some of the genetical myths. I read this as a schoolboy and found it unconvincing and hard to follow, although its liberal message was clear enough. Re-reading it recently showed why: Ashley Montagu had tried, nobly, to make bricks without straw. The information needed to understand our own evolution was simply not available at the time and there seemed little prospect that it ever would be. Human genetics had moved from a series of malign to an equivalent series of pious opinions.

Now everything has been transformed. Humans, far from being the great unknowns of the genetical world, have become its workhorse. More is known about the geographical patterns of genes in the people of the world than about those of any other creature (snails, incidentally, still come second). By about the year 2000 we should have the complete sequence of the three thousand million letters in the DNA alphabet which go to make up a human being. Enough of the genetic message has already been read to make it obvious that the instructions are far more complicated than was ever thought. This knowledge is of more than merely scientific interest. Two out of every three people reading these words will die for reasons connected with the genes they carry; and the new genetics is giving the hope (and for the moment it is only that) of curing inherited disease. We are also beginning to understand what sex really means, why we age and die, and how nature and nurture combine to make us what we are.

Most of all, biology has transformed views of our place in nature. At last it is becoming clear how humans are related to other animals and when and where they first appeared on earth. The idea that life was created—rather than having evolved—is now open to scientific test, and although creationism is still supported by millions of Americans, the test proves it wrong. The study of inheritance is giving new life to the theory of evolution. Most people believe that they descend from simpler predecessors but would be hard put to say why. As Thomas Henry Huxley, Darwin's great Victorian protagonist, said of the idea of evolution: "It is the customary fate of new truths to begin as heresies and to end as superstitions." Genetics has saved Darwinism from this fate. It has killed many superstitions about ourselves. At last there is a real understanding of race, and the ancient and disreputable idea that the peoples of the world are divided into biologically distinct units has gone forever. Separatism may have gained a new popularity among racial groups anxious to assert an identity of their own, but they cannot call on genetics to support their views.

It is the essence of all scientific theories that they cannot resolve everything. Science cannot answer the questions that philosophers—or children—ask: why are we here, what is the point of being alive, how ought we to behave? Genetics has almost nothing to say about what makes people more than just machines driven by biology, about what makes us human. These questions may be interesting, but scientists are no more qualified to comment on them than is anyone else. In its early days, human genetics suffered greatly from its high opinion of itself. It failed to understand its own limits. Knowledge has brought humility to genetics as to other sciences; but the new awareness which genetics brings will also raise social and ethical problems which have as yet scarcely been addressed.

This book is about what genetics can—and cannot—tell us about ourselves. Its title, *The Language of Genes*, points to the analogy upon which it turns, the parallels between biological evolution and the evolution of language.

Genetics is itself a language, a set of inherited instructions passed from generation to generation. It has a vocabulary—the genes themselves—a grammar, the way in which the inherited information is arranged, and a literature, the thousands of instructions needed to make a human being. The language is based on the DNA molecule, the famous double helix, which has become the icon of the twentieth century. Both languages and genes evolve. Each generation there are mistakes in transmission and, in time, enough differences accumulate to produce a new language—or a new form of life. Just as the living tongues of the world and their literary relics reveal a great deal about their extinct ancestors, genes and fossils provide an insight into the biological past. We are beginning to learn to read the language of the genes and it is saying some startling things about our history, our present condition and even our future.

The book emerged from my series of Reith Lectures given on BBC Radio in late 1991. The lectures are named for the famously humorless

John Reith, who was appointed as its first director-general soon after joining the British Broadcasting Corporation in 1922. They began with the philosopher Bertrand Russell in 1948 (and, some would argue, have gone downhill ever since). I would not dream of comparing myself with my illustrious predecessors but I hope that the lectures—and the book —can stand on the merits of their subject, the most fascinating in modern science. The BBC is under attack by those intent on becoming rich by dismembering its corpse. Perhaps my lectures in their small way helped to show that it can still fulfill its obligations, set forth by its founder, to instruct, inform and entertain. The last might seem an unexpected word to use for a series about science, but it is justified by the number of eccentrics and fools who have graced and disgraced the history of human genetics, and who appear sporadically in these pages in the hope of enlivening an otherwise bald narrative.

I would like to thank my producer at the BBC, Deborah Cohen, who did much to turn my ramblings into broadcast form. Cathy Abbott, Lesley Aiello, Sam Berry, David Hopkinson, Tom Jones, David Leibel, Ian Lush, Andrew Pomiankowski and Jenny Sadler read all or part of an early draft of the book and gave valuable advice, not all of which I took. My friend Norma Percy allowed me to work late when writing it without noticing as she was in her own office throughout. Much of the manuscript was written in the French village of Mas Canet, and I thank in particular M. Robert Rigaud for his hospitality. Finally, I should express my gratitude to the various government funding bodies whose determined refusal to support my work led me—as so many others—to abandon research, in my case for journalism. Perhaps, in time, human genetics will help to understand what is really happening in the world of snails, so that this temporary episode of reporting, rather than doing, will not be wasted.

JSJ, January 1994

The Fingerprints of History

IN 1902, IN PARIS, a horrible murder was solved by the great French detective Alfonse Bertillon. He used a piece of new technology which struck fear into the heart of the criminal community. Eighty-three years later two young girls were killed just outside the village of Narborough, near Leicester in the English Midlands. Again, the murderer was found through a technical advance, although the machinery involved would have baffled Bertillon. These events link the earliest and the latest developments of human genetics.

The Parisian killer was trapped because he left a fingerprint at the scene of the crime. For the first time, this was used in evidence as a statement of identity. The idea came from ancient Japan, where a finger pressed into the clay identified the potter. The Leicestershire murderer was caught in the same way. A new test looked for individual differences in genetic material left at the place where the murder had taken place. This "DNA fingerprint" was as much a statement of personal uniqueness as Bertillon's crucial clue or the Japanese potter's mark. As usual, life was more complex than science: the killer, a baker called Colin Pitchfork, was caught only after DNA fingerprints had eliminated a young man who had made a false confession and after Pitchfork had persuaded a friend to give a fraudulent blood sample under his name.

The idea of using fingerprints to trace criminals came from Charles Darwin's cousin, Francis Galton. He founded the laboratory in which I now work at University College London, the first human genetics institute in the world. Every day I walk past a collection of relics of his life. They include some rows of seeds showing similarities between parents and offspring, an old copy of *The Times* and a brass counting gadget which can be hidden in the palm of the hand. Each is a reminder of Galton. As well as revolutionizing detective work he was the first person to publish a weather map and the only one to have made a beauty map of Britain, based on secretly grading the local women on a scale of one to five (the low point, incidentally, being in Aberdeen).

His biography reveals an unremitting eccentricity, well illustrated by the titles of a dozen of his three hundred scientific papers: *On spectacles for divers; Statistical inquiries into the efficacy of prayer; Nuts and men; The average flush of excitement; Visions of sane persons; Pedigree moths; Arithmetic by smell; Three generations of lunatic cats; Strawberry cure for gout; Cutting a round cake on scientific principles; Good and bad temper in English families;* and *The relative sensitivity of men and women at the nape of the neck.* Galton traveled much in Africa, regarding the natives with some contempt and measuring the buttocks of the women using a sextant and the principles of surveying.

Galton introduced the idea that human attributes are coded into the unique inheritance which everyone receives from their ancestors. His work led, indirectly, to today's explosion in human genetics. He was particularly interested in the inheritance of genius (a class within which he placed himself). In his 1869 book *Hereditary Genius* Galton investigated the ancestry of distinguished people and found a tendency for "genius," as he called it, to crop up again and again in the same family. This, he suggested, showed that ability was inborn and not acquired. Galton was the first to try to establish patterns of human inheritance by identifying clearly defined attributes—such as becoming (or failing to become) a judge—rather than speculating about vague qualities such as musicality or fecklessness.

Galton and his followers would be astonished at what biology can now do. Although it is scarcely any closer to understanding traits such as genius (and reputable scientists hardly concern themselves with them), genetics is in a phase of explosive growth. DNA fingerprints are routinely used in criminal cases, and there are plans—much contested by civil libertarians—to set up data bases of such information in the United States and in Europe. The same kind of variation is used in medicine. Personal DNA patterns can track down damaged genes and allow parents to choose whether to risk having a child with an inborn disease. Nearly five thousand such diseases are known and if we include, as we should, all ailments (such as cancer and heart disease) which have an inherited component, then most people die because of the genes they carry.

Biology does more than reveal our individual fate. The genes show that humans share much of their heritage with other creatures. As Galton himself discovered (and illustrated by pasting the appropriate impression near that made by Gladstone, the prime minister), chimpanzees have fingerprints. Now we know that much of their DNA is identical to our own. All this suggests that we and the chimps are close relatives. Many of our genes are shared with creatures as different as mice, bananas and bacteria.

Genetics is the key to the past. Every human gene must have an ancestor. This means that patterns of inherited variation can be used to piece together a picture of history more complete than from any other source. Each gene is a message from our forebears and together they contain the whole story of human evolution. Everyone is a living fossil, carrying within themselves a record which goes back to the beginnings of humanity and far beyond. The most famous line in *The Origin of Species* expresses the hope that "light may be shed on man and his origins." Darwin's hint that humans share a common descent with every other creature is now accepted by all biologists, largely because of the evidence of the DNA.

Darwin described evolution, the appearance of new forms of life by the alteration of those already present, as "descent with modification." The same phrase could be used of language. Because of mutation, the language of the genes is liable to be garbled during transmission. As a boy, I was greatly amused by the tale of the order given down the line of command to soldiers in the trenches. "Send reinforcements, we're going to advance" changed to "Send three and fourpence, we're going to a dance" as it passed from man to man. This simple tale illustrates how accidents in copying an inherited message can lead to change.

This book is about inheritance, and about evolution: about the clues of our past, present and future which each of us contains. The language of the genes has a simple alphabet, not with twenty-six letters, but just four. These are the four different DNA bases—adenine, guanine, cytosine and thymine (A, G, C and T for short). The bases are arranged in words of three letters such as CGA or TGG. Most of the words code for different amino acids, which themselves are joined together to make proteins, the building blocks of the body.

Just how economical the language of inheritance is can be illustrated with a rather odd quotation from a book called *Gadsby*, written in 1939 by the American author Ernest Wright: "I am going to show you how a bunch of bright young folks did find a champion, a man with boys and girls of his own, a man of so dominating and happy individuality that youth was drawn to him as a fly to a sugar bowl." This sounds rather peculiar, as does the rest of the fifty-thousand-word book, and it is. The quotation, and the whole book, does not contain the letter "e." It is possible to write a meaningful sentence with twenty-five letters instead of twenty-six, but only just. Life manages with a mere four.

Although the inherited vocabulary is simple its message is very long. Each cell in the body contains about six feet of DNA. A useless but amusing fact is that if all the DNA in all the cells in a single human being were stretched out it would reach to the moon and back eight thousand times. There is now a scheme, the Human Genome Project, to read the whole of its three thousand million letters and to publish what may be the most boring book ever written; the equivalent of a dozen or so copies of the *Encyclopaedia Britannica*. There is much disagreement about how to set about reading the message and even about whether it is worth doing at all. It probably is. The British Admiralty sent the *Beagle* to South America with Darwin on board not because they were interested in evolution but because they knew that the first step to understanding (and, with luck, controlling) the world was to make a map of it. The same is true of the genes. To make this map will be expensive—about as much as a single Trident nuclear submarine. The task will be stupefyingly tedious for those who have to do the work, but, soon after the end of the century, someone will publish the inherited lexicon of a human being. To be more precise, there will be a map of a sort of Mr. Average—the chart is, of course, of a male—as the information will come from short bits of DNA from dozens of different people.

Powerful ideas, such as those of inheritance and evolution, attract myths. Impressed by his studies of genius, Galton founded the science (if that is the right word) of eugenics. Its main aim was "to check the birth rate of the Unfit and improve the race by furthering the productivity of the fit by early marriages of the best stock." By so doing he led the newly established field of human genetics into a blind alley from which it did not emerge for half a century. At his death, he left £45,000 to found the Laboratory of National Eugenics at University College London and, in fine Victorian tradition, £200 to his servant who had worked for him for forty years. His laboratory soon changed its name to the Galton Laboratory to escape from the eugenical taint. What became of his servant is not recorded.

Galton's eugenical ideas and Darwin's evolutionary insights had a pervasive effect on the intellectual and political life of the twentieth century. They influenced both left and right, liberal and reactionary. They continue, explicitly or implicitly, to do so. Many apparently disparate figures trace their ideas to *The Origin of Species* and to *Hereditary Genius*. All are united by one thing: a belief in biology as destiny, in the power of genes over the people who carry them.

The most famous monument in Highgate Cemetery in London, a couple of miles north of today's Galton Laboratory, is that of Karl Marx. Its inscription is well known: "Philosophers have only interpreted the world. The point, however, is to change it." Darwinism was soon used to attempt to make just such a change. The founder of Social Darwinism, the idea that poverty and wealth are inevitable as they represent the biological rules which govern society, was the philosopher Herbert Spencer. He is buried just across the path from Marx. In his day, Spencer was famous. His *Times* obituary claimed that "England has lost the most widely celebrated and influential of her sons." Now he is remembered only for that neatly circular phrase "the survival of the fittest" and for inventing the word "evolution." He wrote with a true philosopher's clarity: "Evolution is an integration of matter and concomitant dissipation of motion; during which the matter passes from an indefinite, incoherent homogeneity, to a definite, coherent heterogeneity, through continuous differentiations and integrations." He was parodied by a mathematical contemporary: "A change from a nohowish, untalkaboutable all-alikeness to a somehowish and in general talkaboutable not-all-alikeness by continuous somethingelsifications and sticktogetherations."

Spencer used *The Origin of Species* as a rationale for the excesses of nineteenth-century capitalism. The steel magnate Andrew Carnegie was impressed by the idea that evolution excuses injustice. He invited Herbert Spencer to Pittsburgh. The philosopher's response to seeing his theories worked out in steel and concrete was that "Six months' residence here would justify suicide."

Galton, too, was all in favor of interfering with human evolution. He supported the idea of breeding from the best and sterilizing those whose inheritance did not meet with his approval. The eugenics movement combined a touching concern for the unborn with a brutal rejection of the rights of the living—a combination not unknown today.

Galton's main interest in inheritance was as a means of judging the quality of his fellow citizens and of averting the imminent decline of the human race. He claimed that families of "genius" had fewer children than most and was concerned about what this meant for future generations. Perhaps his own childless state helped to explain his anxiety.

Many of the early eugenicists shared some highly heritable attributes: wealth, education and social position. Francis Galton gained his considerable affluence from his (rather unusual) family of Quaker gunmakers. Much of the agenda of their movement was the survival of the richest. Other eugenicists were on the left. They felt that if economies could be planned, then so could genes. George Bernard Shaw, at a meeting attended by Galton in the last few years of his life, claimed that "Men and women select their wives and husbands far less carefully than they select their cashiers and cooks." Later, he wrote that "Extermination must be put on a scientific basis if it is ever to be carried out humanely and apologetically as well as thoroughly." Shaw was, no doubt, playing his role as Bad Boy to the Gentry, but subsequent events made his play-acting seem even less amusing than it did at the time.

The eugenics movement had a greater practical effect in the United States than in Galton's native country. In 1898, Charles Davenport, then professor of evolutionary biology at Harvard, was appointed as Director of the Cold Spring Harbor Laboratory on Long Island Sound. Initially, the laboratory concentrated on the study of "the normal variation of the animals in the harbor, lakes and woods, and the production of abnormalities." It carried out some of the most important work in early-twentieth-century biology.

Soon, though, Mrs. E. H. Harriman, widow of the railway millionaire, decided to devote part of her late husband's fortune to the study of human improvement. The Eugenics Record Office was built next to the original laboratory. It employed more than two hundred field workers, who were sent out to collect pedigrees. Their 750,000 genetic records included studies of inherited disease and of color blindness; but also recorded the inheritance of characteristics such as shyness, pauperism, nomadism, moral control and shiftlessness.

Davenport's work had an important effect on American society. In the early part of this century there were eugenical clubs with prizes for the fittest families, the name "Eugene"—well-born—entered the English language and, more seriously, medicine became concerned about whether its duty to the future outweighed the interests of some of those alive today. Twenty-five thousand Americans were sterilized because they might pass feeble-mindedness or criminality to future generations. One judge, saying that "three generations of imbeciles are enough," compared sterilization with vaccination. The common good, he said, overrode individual rights.

Another political leader had similar views. "The unnatural and increasingly rapid growth of the feeble-minded and insane classes, coupled as it is with steady restriction among all the thrifty, energetic and superior stocks constitutes a national and race danger which it is impossible to exaggerate. I feel that the source from which the stream of madness is fed should be cut off and sealed off before another year has passed." Such were the words of Winston Churchill when he was Home Secretary in 1910. His beliefs were felt to be so inflammatory by succeeding British governments that they were not made public until 1992.

One of Galton's followers was the German embryologist Ernst Haeckel, who was enormously impressed by *The Origin of Species*. Haeckel was a keen supporter of evolution. He came up with the idea (which later influenced Sigmund Freud) that every animal relived its evolutionary past during its embryonic development. Haeckel was more than a biologist. He was an enthusiast for social change. His reading of Galton and Darwin and his belief in inheritance as fate led him to found the Monist League, which had thousands of members in Germany before the First World War. It argued for the application of biological rules to society and, in particular, for the survival of some races—supposedly those with the finest biological heritage—at the expense of others. Haeckel claimed that "the whole history of nations can be explained by means of natural selection" and that "social rules are the natural laws of heredity and adaptation." He felt that the evolutionary destiny of the Germans was to overcome inferior peoples: "The Germans have deviated furthest from the common form of ape-like men . . . The lower races are psychologically nearer to the animals than to civilized Europeans. We must, therefore, assign a totally different value to their lives." It is no coincidence that Hitler's biography, *Mein Kampf*, "My Struggle," has as its title part of the translated Darwinian phrase, the struggle for existence.

In 1900 the arms manufacturer Krupp (whose descendants made fortunes during the Second World War) offered a prize of thirty thousand Reichsmarks for the best essay on "What Can the Theory of

Evolution Tell Us about Domestic Political Development and the Legislation of the State?" There were sixty entries. His fellow industrialist Siemens expressed concern that "the best of human heredity will be swamped by a mass of inferior types." The first German eugenic sterilization, though, was carried out by a socialist doctor (although he was one who claimed that trade union leaders were more likely to be blond than were those who followed them).

While imprisoned after the Beer Hall Putsch, Hitler read the standard German text on human genetics, *The Principles of Human Heredity and Race Hygiene*, by Eugene Fischer. Fischer was the director of the Berlin Institute for Anthropology, Human Heredity and Eugenics. One of his assistants, Joseph Mengele, later achieved a certain notoriety for trying to put Galtonian ideas into practice. Fischer's book contained a chilling phrase: "The question of the quality of our hereditary endowment"—it said—"is a hundred times more important than the dispute over capitalism or socialism."

His thoughts were echoed in *Mein Kampf:* "Whoever is not bodily and spiritually healthy and worthy shall not have the right to pass on his suffering in the body of his children." Hitler took the idea to its horrifying conclusion by destroying the people he saw as less favored and attempting to breed from the best. The task of improving hereditary quality was taken seriously. There were four hundred thousand sterilizations of those deemed unworthy to pass on their genes, sometimes by the secret use of X-rays as the victims were filling out forms. Those in charge of the program in Hamburg estimated that one fifth of the population deserved to be treated in this way.

By 1936 the German Society for Race Hygiene had more than sixty branches and doctorates in racial science were offered at several German universities. The training emphasized the view that certain peoples were inferior because of their inheritance. Half of those at the Wannsee Conference, which decided on the final solution of the Jewish problem, had doctorates, mainly in anthropology. Many of them justified their crimes on scientific grounds. The eugenics movement in Germany was strongly opposed to abortion (except, of course, of the biologically unfit) and imposed stiff penalties—up to ten years in prison—on any doctor rash enough to carry it out. By this and other measures the Nazi eugenicists succeeded in reversing the "birth strike" of the 1930s. The number of children born to women of approved stock went up by a fifth. The conjunction of extreme right-wing views, an obsession with racial purity and the anti-abortion movement has its echoes in the United States today. Nazi concern for the purity of German blood reached absurd lengths. One unfortunate member of the National Socialist Party received a transfusion from a Jew after he had been in a road accident. He was brought before a disciplinary court to see if he should be excluded from the Party. Fortunately, though, the donor had fought in the First World War, so that his Jewish red cells were—just about—acceptable.

The disaster of the Nazi experiment effectively ended the eugenics movement, at least in its primitive form. Its blemished past means that human genetics is marked by the fingerprints of its own history. Sometimes it seems to find it difficult to wipe them off. They must not be forgotten, particularly as the subject is now, for the first time, in a position to control the biological future.

Galton and his followers felt free to invent a science which accorded with their own prejudices. They believed that our duty to our genes outweighs that to those who bear them. They were filled with extraordinary self-assurance. Their views were taken seriously although in retrospect it is obvious that they knew almost nothing about human inheritance. Today's new knowledge is likely to prove as controversial as the old ignorance. One thing has changed. Disputes among modern biologists are not about the vague general issues which obsessed their predecessors. Instead they concern themselves with the fate of individuals rather than of all humanity. Human genetics has become a science and, as such, has narrowed its horizons.

Nevertheless, genetics raises ethical issues which are not going to go away. By diagnosing defective genes before birth it is possible to shift the balance between birth and natural abortion to reduce the number of damaged children. This raises passions, from those who feel—in spite of the high natural wastage of fertilized eggs—that all fetuses are sacred, to others who consider that knowingly to pass on a damaged gene is equivalent to child abuse. Genetics also presents a more subtle and more universal dilemma—the problem of knowledge. Soon, it will tell many of us how and when we are likely to die. Already, it is possible to diagnose at birth genes which will kill in childhood, youth or early middle age. More will soon be found. Will people really want to know that they are at risk of a disease about which they can do nothing? Many diseases show their effects only if a child inherits a copy of the damaged DNA from each parent. Everyone is likely to pass on a single copy of at least one such gene. Will this help in choosing a partner or deciding whether to have children? The greatest dilemma of all will be that of being aware of our own fate or that of our offspring.

The ethical problems which the new biology raises can be illustrated with a very old piece of discrimination based on perceptions of inborn quality. There has always been prejudice against certain genes, those carried on the chromosomes which determine sex. Women have two "X" chromosomes, men a single X chromosome and a much smaller "Y" chromosome. This means that all eggs have an X but that sperm are of two kinds, X or Y. At fertilization, both XY males and XX females are produced in equal number. Sex is as much a product of genes as are blood groups.

The way in which we judge the value of these genes helps to understand how people make biological decisions and how moral judgments depend on circumstances. Sometimes, Y chromosomes seem to be worth less than Xs. When it comes to wars, murders and executions males have always been more acceptable victims than females. But the balance can shift. Many parents express a preference for sons, especially as a firstborn. Some actually try to achieve them. The recipes vary from the heroic to the hopeful. In ancient Greece, tying off the left testicle was said to do the job, while medieval husbands drank wine and lion's blood before copulating under a full moon. Less drastic —but equally ineffective—methods included mating only in a north wind or hanging one's underpants on the right side of the bed.

Being a gender vendor has always been an easy way to make money. It has, after all, a guaranteed fifty percent success rate. Methods promoted today vary from using baking soda or vinegar at the appropriate moment (to take advantage of a supposed difference in the resistance of X- and Y-bearing sperm to acids and alkalies) to mating at particular times of the female cycle (with an appropriately expensive watch as part of the recipe). A diet high or low in salt is also said to make a difference. Most of these recipes are useless and a number of those who sell them have been prosecuted for fraud.

Now genetics means that fraud is out of date and that ethics is becoming a problem. There are several ways to make an effective choice of sex. One is to separate X and Y sperm and to fertilize a woman only with the appropriate type. Since Louise Brown in 1978, hundreds of children have been born by in vitro fertilization, adding sperm to egg in a test tube. A single cell can be taken from the developing embryo—usually when there are only eight cells altogether— and its sex determined. Only those of the desired sex are implanted into the mother. This technique (known, rather excruciatingly, as BABI, for Blastomere Analysis Before Implantation) has led to the birth of more than two hundred babies. It raises questions of its own. What, for example, should be done with spare embryos found to be normal?

In Eskimo society fifty years ago there were twice as many boys as girls, because girls were killed at birth or allowed to starve when food was short. In parts of modern China, too, the one-child policy has produced an excess of boys, with dark hints of infanticide. The Indian government recently shut down clinics which chose the sex of a baby by looking at the chromosomes of the fetus—and aborting those with two Xs. More than two thousand pregnancies a year were ended for this reason in Bombay alone. The main reason was the need for large dowries when daughters were married off. The clinics advertised with slogans such as "Spend six hundred rupees now, save fifty thousand later." The effect is not trivial. India is one of the few countries of the world where there are fewer females than males—four girls to five boys in some states—and, because of infanticide and selective abortion, there is an overall deficit of Indian girls and women equivalent to a fifth of the female population of the United States.

All these methods of choosing sex interfere with genes. Their acceptability varies from the reasonably uncontentious choice of sperm to a crime where infanticide is concerned. Where to draw the line depends on one's own social, political or religious background. Most readers of this book would, I imagine, see the possibility of terminating a pregnancy just because it is the wrong sex as being ethically unacceptable. They might worry less about choosing X or Y sperm. Many do not like the idea of deciding that boys are worth more than girls, or vice versa. They object to the notion that life or death should depend on a child's biological merit.

There are more serious issues than parental self-indulgence when it comes to choosing a child's sex. Sometimes the judgment is a matter of life and death. Many inherited diseases are carried on the X chromosome. In girls, an abnormal X is usually masked by a normal copy of the same chromosome. Boys do not have this option, as they have only a single X. For this reason, sex-linked abnormalities, as they are known, are much more common in boys than in girls. They can be very distressing. Sex-linked muscular dystrophy is a wasting disease which leads to weakness and death in a child's teens or twenties. A couple who have had a son with muscular dystrophy can scarcely be blamed for wanting to make sure that none of their future children has the disease. They want to make a decision about the biological quality of their offspring and few will criticize them for doing so. An understanding of how genetics works means that their personal ethical balance has begun to shift.

If a couple has an affected child they immediately know that the mother carries the gene. The chance of a second son having the disease is hence far greater than before. It is still less than half, so that if all male pregnancies are terminated there is a real possibility of losing a normal boy. Most people who dislike the idea of choosing a child's sex through—say—using X-bearing sperm might change their minds in these circumstances. Many others would accept the idea of terminating all pregnancies which would produce a son if there is a real chance of muscular dystrophy.

Recently, the choices have become more precise. The gene for muscular dystrophy has been tracked down and changes in the DNA can show whether a fetus is carrying it. There are now more than two hundred centers worldwide which use this test. But the method is not yet perfect. There are many ways in which the gene can go wrong and not all of them show up. There is hence a danger that a fetus that appears normal may in fact be carrying the gene, complicating the parents' decision as to whether to continue with a pregnancy. There is also a certain hazard in sampling fetal tissues. This is getting smaller as technology improves (for example by testing the tissue surrounding the very early embryo, or by looking for fetal cells in the mother's blood) but the risks of making the test must themselves be weighed in the moral balance.

It seems inevitable that parents' decisions about the future of their unborn child will increasingly be influenced by estimates of risk and of quality: by whether the rights of a fetus depend on the genes it carries. Ethics will depend more and more on circumstances. Genetical judgments will be needed as we learn more about our inheritance. They are not scientific decisions, but depend on the society and the individuals who make them. Since the debacle of the eugenics movement there has been an understandable reluctance even to consider the idea of making choices about individual rights based on differences in inherited merit; but the new knowledge means that such choices are unavoidable.

After the Second World War genetics had—it seemed—at last begun to accept its own limits and, by so doing, to escape its confines as the haunt of the politically obsessed. Most of those in the field today are cautious about claims that the essence of humanity lies in its DNA. Although it can certainly divulge extraordinary things about ourselves genetics is one of the few sciences which has reduced its expectations.

Science owes most of its success to the fact that it is reductionist: to understand something, it must be broken down into its component parts. The approach certainly works well in genetics as far as it goes, but it only goes so far. Where it does not work can be seen in a phrase notorious in British politics, the former Prime Minister Mrs. Thatcher's statement that "There is no such thing as society, there are only individuals." The failures of this philosophy are all around us. To say, with Galton and his successors, "There are no people, there are only genes" is to fall into the same trap.

In spite of the lessons of the past there has been a resurgence of the dangerous and antique myth that biology can explain everything. Some psychologists and politicians have again begun to claim that we are controlled by our inheritance. They promote a kind of biological fatalism. Humanity, they say, is driven by its inheritance, and biology is a sort of original sin. According to this view, those who do not succeed are victims of their own heritage. Their predicament is due to their own weakness and has nothing to do with the rest of us. Such *nouvelle* Calvinism suggests that as human life was programmed long ago there is no point in trying to change it—which is convenient for those who like things the way they are.

In some parts of the world the new eugenics is quite overt. Lee Kuan Yew, Prime Minister of Singapore, was concerned by the low birth rate of educated women and produced a financial package designed to persuade them to improve the quality of future generations by having more children. The Chinese *People's Daily* was franker in its views. In 1988 it reported a scheme to ban the marriage of those with mental disease unless they were sterilized with a robust simplification of Mendelism: "Idiots give birth to idiots!"

Sometimes the eugenical message is justified on financial grounds. At the Sesquicentennial Exhibition in Philadelphia in 1926 the American Eugenics Society had a board with flashing lights counting up the $100 per second allegedly spent on people with "bad heredity." Sixty years later, one of the proponents of the plan to sequence the human genome claimed that the project would pay for itself just by "curing" the problem of schizophrenia—and by curing, of course, he meant terminating pregnancies shown to be carrying the as yet hypothetical and undiscovered gene for the disease. The 1930s were a period of financial squeeze for health care. The 1990s, too, are a time when the state is attempting to cut the amount spent on medicine. There is a fresh danger that genetics will be used as an excuse to discriminate against the handicapped in order to save money.

In medieval Japan, the science of dactylomancy—judging personality from fingerprints—had it that people with complex patterns were likely to be good craftsmen, those with many loops lacked perseverance, while those whose fingers carried an arched pattern were crude characters without mercy. Human genetics has escaped from its dactylomantic beginnings. The more we learn about inheritance the more it seems that there is to know. The shadow of eugenics has not yet disappeared but it is much fainter than it was. Now that genetics has matured as a subject it is beginning to reveal an extraordinary picture of who we are, what we were, and what we may become. This book is about what that picture contains.

The Language of Genes

1 | A Message from Our Ancestors

THE OLD ENGLISH FAMILY the Temple-Nugent-Bridges-Chandos-Grenvilles—now, alas, extinct—was justifiably proud of its heritage. Although the family name had a mere five barrels, the coat of arms had seven hundred and nineteen symbols, each showing a link to another aristocratic lineage. The family's ancestry was preserved, the world could see and appreciate it and those skilled enough to read the message on the shield could learn a great deal about their history.

Because wealth and social position depend so much on descent the rich were the first geneticists. Vague statements of historical importance were not enough. They needed—and awarded themselves—concrete symbols of who they were and from whom they were descended which could persist even when their carriers were long dead. Such statements of shared inheritance have endured for hundreds of years. The Lion of the Hebrew Tribe of Judah was, until recently, still the symbol of the Emperor of Ethiopia. Those of England trace directly back to the lions awarded to Geoffroy Plantagenet in 1127. The fetish for ancestry means that royal families play an important part in genetics—and one enthusiast has even traced 262,142 alleged ancestors of Prince Charles.

Even nations more rational than those of the Old World retain an obsession for inheritance. The Daughters of the American Revolution pursue with zeal those who claim a descent to which they are not entitled. Ananias Dare—whose daughter Virginia was the first child born in America of English parentage—received a coat of arms from the Royal College of Heralds in 1588, the earliest given to an American. The development of heraldry was cut off by the Revolution and the last grant of arms was made to the magnificently named Andrew Pepperell Sparhawk in 1775. Even George Washington corresponded with the Heralds in the hope of establishing a link with the Washingtons of Northamptonshire, whose five-pointed stars (used illegally by him as a bookplate) were granted in 1819.

Heraldic symbols are reports from our forefathers. They exist because it was realized long ago that only when the past is preserved in discrete form does it make sense. For much of history—from the pharaohs to the hundred thousand Victorian bourgeoisie who filled Highgate Cemetery in London with monuments (including a copy of a large section of the pharaonic city of Luxor)—family wealth was dissipated on funerary ornaments to remind unborn generations from whom they descend. As well as its Galtoniana, University College London contains another eccentric object. This is the stuffed body of the philosopher Jeremy Bentham (who was associated with the College at its foundation). Bentham hoped to start a fashion for such "auto-icons" in the hope of reducing the cost of less organic monuments to the deceased. The fashion did not catch on, although the continuing popularity of Bentham's corpse with visitors suggests that it ought to have done.

The pride of Bentham and the Temple-Nugent-Bridges-Chandos-Grenvilles in their heritage would now be greeted, mainly, with derision. Harold Wilson, the British Prime Minister of the 1960s, did as much when he mocked his predecessor, Lord Home, for being the Seventeenth Earl of that name. Lord Home mildly and accurately deflected the jest by pointing out that his critic was presumably the seventeenth Mr. Wilson. He was making a valid biological point: that while only a chosen few preserve their heritage in an ostentatious way, every family, aristocratic or not, inherits in their genes a record of who their ancestors were and where they came from.

I can trace my own back only to my great-great-grandfather, who lived in the Welsh village of New Quay where I myself spent some of my youth. Some people can use family history to follow up their lineage for far longer. However, everyone can decipher much more of their ancestral record by looking at the biological heritage preserved in the DNA.

Sometimes the way in which biology is a guide to the past is obvious. In some aristocratic families coats of arms travel down the generations together with genes. The Hapsburg name and its crest, a double-headed eagle, lived through a thousand years of European history. They started with the (no doubt well nicknamed) Guntram the Rich of Hapsburg—the Hawk's Castle—in 950. Those inheriting the name and the crest were liable to get something else, a gene for a protruding lower jaw which became known as the Hapsburg lip. The famous lip can be seen in the Holy Roman Emperor of 1450 and was still prominent in the Spanish royal family a century ago. The ancestry of the Hapsburgs is written on their faces as much as on their shield.

We notice the Hapsburgs because they are different. They have a slight deformity which distinguishes them from other families. Thousands of inherited abnormalities are now known. Some are damaging and do not last for long. Others are milder and, like the royal lip, can be used to trace shared descent for hundreds of years. A form of juvenile blindness—hereditary glaucoma—is found in France. Painstaking searches of parish records show that most of the cases are descended from a single couple who lived in the village of Wierr-Effroy near Calais in the fifteenth century. Even today pilgrims pray in the village church of St. Godeleine, which contains a cistern whose waters are believed to cure blindness. Thirty thousand descendants of this couple have been traced and for many the diagnosis of the disease was their first clue about where their ancestors came from and who their relatives are.

The gene went with French emigrants to Canada and—probably—to Louisiana. For Quebecois and Cajuns, too, it is a link with their European ancestors. The early history of the Americas preserves other genetic connections with the past. The Church of Latter-Day Saints has, in Salt Lake City, one of the finest collections of pedigrees in the world. It shows the family ties between ten thousand Mormon pioneers and a million and a half of their living descendants. Sometimes it is possible to work out from nineteenth-century medical records just what those pioneers died of. If a modern Mormon woman has an ancestor among the first settlers who suffered from, for example, breast cancer, she herself is at a relatively higher risk of this disease—good evidence that genes are involved and that they can exert their malign effect on people born two centuries apart.

Human genetics was for most of its history more or less restricted to studying pedigrees which stood out because they contained an abnormality. This limited its ability to trace patterns of descent to those few families—like the Hapsburgs—who appeared to deviate from the perfect form. Biology has now shown that this perfect form does not exist.

Instead there is a huge amount of inherited variation. Thousands of inherited characters—perfectly normal diversity, not diseases—distinguish each person. There is so much variety that everyone alive today is different, not only from everyone else, but from everyone who ever has lived or ever will live. The mass of diversity can be used to look at patterns of shared ancestry in any family, aristocratic or plebeian; healthy or ill. Because all modern genes are copies of those in earlier generations each can be used as a message from the past. They bring clues from the beginnings of humanity more than a hundred thousand years ago and from the origin of life three thousand million years before that.

Most of modern genetics is nothing more than a search for variation. Some of the differences can be seen with the naked eye. Others need the most sophisticated methods of molecular biology. As a sample of how different each individual is—information which is needed to work out how closely related everyone might be—we can glance beneath the way we look (Hapsburg lip or not) to ask about variation in how we sense the world and how the world perceives us.

Obviously, people look different from each other. The inheritance of appearance is not particularly simple. Eye color depends first on whether there is any pigment present. If there is none the eye is pale blue. Other colors vary in the amounts of pigments controlled by several distinct genes. Comparing eye colors is, perhaps fortunately, not a dependable way of working out who is related to whom. The inheritance of hair color is also rather complicated. Apart from very blond or very red hair, the rest of the range is genetically confused and also involves age and exposure to the sun. The range of colors of the children of African and European parents suggests that around half a dozen genes control their differences in skin color but not much is known of the details.

Even a trivial test shows that individuals differ in many other ways. Stick your tongue out. Can you roll it into a tube? About half those of European descent can and half cannot. Clasp your hands together. Which thumb is on top? Again, about half the population folds the left thumb above the right and about half do it the other way. These characteristics certainly run in families but the details of their inheritance—like those of physical appearance—are uncertain.

People differ from each other not only in the way the world sees them, but how they see it. A few are color-blind. They lack a receptor for red, green or blue light. All three are needed to perceive the full range of color. The absence of one (usually that for red or green) is a mild disability. It may have made some difference when gathering food in ancient times. The genes involved have now been tracked down. During the search it emerged that color blindness is just the extreme end of a system of normal variation. When asked to mix red and green light until they match a standard orange color, people divide into two groups which differ in the hue of the red light which they choose. This is because there are two distinct receptors for red. Each differs in a single change in the DNA. About sixty percent of Europeans have one form, forty percent the other. Both groups are normal (in the sense that they are aware of no handicap) but one sees the world through slightly more rose-tinted spectacles than does the other. The contrast is small but noticeable. If two men with different red receptors chose jacket and trousers for a Santa Claus costume there would be a perceptible clash between upper and lower halves.

There are other subtle differences in perception of the outside world. In the 1930s, an American manufacturer of ice trays was surprised to receive complaints from his customers, who claimed that their ice tasted bitter. This baffled the unfortunate entrepreneur as the ice tasted just like ice to him. It turned out that there are inherited differences in the ability to taste a chemical used in the manufacturing process. To some, a trace of this chemical, PTC, is intolerably bitter, while to others a concentration a thousand times greater has no taste at all. The difference depends on just one gene which exists in two forms —taster and non-taster. When I was a student it was regarded as amusing to make tea containing PTC and to observe the bafflement of those who could drink it and those who could not. Now, unfortunately, students have more sense.

Many creatures communicate with each other as much by scent as by vision. Female mice can tell from smell not only who a male is, but how close a relative he might be. There is an intriguing hint that humans also have an identity based on smell. Police dogs find it more difficult to separate the trails of identical twins (who have all their genes in common) than those of unrelated people. The human species has more scent-producing glands than does any other primate, and perhaps there is a remnant of a system of uniqueness in scents which has lost its importance in a world full of sight.

Variation in the way we look, see, smell and taste is accompanied by inherited diversity in nearly all our attributes. The genes which enable mice to recognize each other by smell are part of a much larger system of identifying outsiders. The threat of infection means that there is a constant conflict with the external world. The immune system determines what should be kept out. It can differentiate "self" from "notself." Having done so, it makes protective antibodies which interact with antigens (chemical clues on a native or foreign molecule) to define whether an unfamiliar substance is acceptable. The immune system produces millions of antibodies, each of which recognizes a single antigen. Cells bear antigens of their own. Like individual appearance, but with far more precision, they differentiate us from all our fellows. Antigens on the surfaces of cells gave the first hint of the mass of hidden genetic variation which all humans possess.

When blood from two people is mixed, it may turn into a sticky mess —which would be fatal in a blood transfusion. This process is controlled by a system of antigens, the blood groups. Only certain combinations can be mixed successfully. There are plenty of different blood group systems. Some are familiar, like ABO and Rhesus. Others, like Duffy and Kell, are less so. Millions of people have been tested. A dozen systems are routinely screened, each with a number of different forms. A mass of diversity is generated by just this small sample of genes. The chances of two Englishmen having the same combination of all twelve blood groups is only about one in three thousand and, because of the mixing of so many genetically distinct peoples in the New World, that for two Americans is even less.

Other statements of individual identity on the surfaces of cells are even more variable. These, too, evolved as an assertion of identity, to ensure immediate recognition of what comes from the outside world and what belongs within. This "histocompatibility system" (which is important in organ transplantation) provides another set of inherited statements of ancestry which can be used to trace relatedness.

Blood groups and the other cell-surface antigens were discovered before molecular biology began. Since then, there has been a technical revolution. Like the stone age revolution a thousand centuries ago it depends on simple tools which can be used in many ways. Now it is possible to compare the DNA of different individuals, either letter by letter, or by asking where and how often phrases or paragraphs are repeated.

The comparison shows that, just as in blood groups, everyone is different. On the average, two people differ in about one DNA letter per thousand. This gives about three million places in the inherited message which differ from person to person. Blood groups show how unlikely it is that two will be the same when only twelve variable systems are used. The chance that they both have the same sequence of letters in the whole DNA alphabet is one in hundreds of billions.

Personal uniqueness itself says something useful: molecular biology has made individuals of us all. Genetics disproves Plato's myth of the absolute, that there exists one ideal form of human being from which there are rare deviations such as those who have an inborn disease.

Inherited variation helps us to understand where we fit in our own family tree, in the pedigree of humankind, and in the living world as a whole. Relatives are more likely to share genes than are unrelated people because they have an ancestor in common. As genes all descend from a predecessor they can be used to test relationship, however distant this might be. The more variants two people share the more closely they are related. The same logic can be used to sort out more remote patterns of affinity, including the shared ancestry of humans and other creatures.

Detective work of this kind is easy when close relatives are involved. There is a macabre plan by the U.S. Army to test the relationship of dead bodies to their previous owners by storing DNA samples from soldiers in the hope of identifying their mutilated corpses after death. DNA can also say a lot about family connections. Before such tests were available, immigration officers faced with applicants for entry often refused to believe that a child was the offspring of the woman who claimed it. Comparing the genes of mother and child nearly always showed that the mother was telling the truth. Society being what it is, the tests are used less now than they once were.

Not all families are what they seem. Attempts to match the genes of parents and offspring in Britain or the United States reveal quite a high incidence of false paternity. Many children have a combination of genes which cannot be generated by combining those of their supposed parents. Usually, they show that the biological father is not the male who is married to the biological mother. In middle-class society about one birth in twenty is of this kind. On a more positive note, the use of DNA tests in a survey of American rape victims who had become pregnant showed that most of the children were in fact fathered by their husbands.

Paternity can be tested from beyond the grave. DNA is tough stuff, which can persist long after the death of its owner. The eugenical enthusiast Joseph Mengele fled to South America after the end of the Second World War. He was—allegedly—sighted on many occasions. In the late 1980s, bones supposed to belong to Mengele were discovered. His son gave a blood sample. Comparison of his genes with the DNA in the remains of his putative father showed that the bones were indeed those of Joseph Mengele, who had been found, but too late for justice. Shared diversity also makes it possible to skip generations in the search for ancestry. During the Argentinian military dictatorship of the 1970s and early 1980s thousands of people disappeared. Most were murdered. Some of the victims were pregnant women who were killed after they had given birth. Many of the children were stolen by military families. When civilian rule was restored after 1983 a group of mothers of the murdered women began to search for their grandchildren. The DNA of the children was matched with that of those who claimed to be their grandparents. The message passed in the genes over two generations enabled more than fifty of the children to be restored to their biological families from those who had snatched them.

Some families have no hope of restoration. Bones dug up in a cellar in Ekaterinburg have long been suspected to be those of the last tsar and his family, shot in 1918. Some of his living relatives (including the Duke of Edinburgh) have given DNA. Comparing the two sets of genes proves that the skeletons are, indeed, the remains of the Romanovs. Intriguingly enough, the skeleton of one young girl imprisoned with the group was missing. A woman known as Anna Anderson claimed for many years to be Anastasia, the daughter of the tsar. She had, she said, escaped execution. Her claim was rejected by a German court; but perhaps the bones give some credence to her tale.

Everyone's individual combination of genes comes from ancestors who died long before their great-grandparents. It is in some ways a genetic coat of arms. Like the shield of the Temple-Nugent-Bridges-Chandos-Grenvilles it contains a record of who the forebears were and to whom they were related. When people move they take their DNA with them, so that making maps of genes in modern humankind does more than just trace descent. Genetics can re-create history.

Sometimes history itself is a clue as to where to start. Alex Haley, in his book *Roots*, used documents on the slave trade to try to find his African ancestors. He found only one, Kunta Kinte by name, who had been taken as a slave from the Gambia in 1767. Patterns of inherited diversity in today's black Americans could have told him much more.

The African slave trade began in the early days of the Roman Empire. By A.D. 800 Arab traders had extended it to Europe, the Middle East and China. In the fifteenth century the Spanish and Portuguese started what became a mass forced migration, initially from the Guinea Coast, modern Mauritania. The migrants reached much of Europe: there were black gondoliers in medieval Venice and by the sixteenth century one person in ten in Lisbon was black. The trade had the support of the Church. A bull of Pope Nicholas V instructed his followers to "attack, subject, and reduce to perpetual slavery the Saracens, Pagans and other enemies of Christ, southward from Cape Bojador and including all the coast of Guinea."

The main slave trade was to the New World. About fifteen million Africans were shipped to the Americas. They came from all over West Africa from Senegal to Angola and were dispersed over much of North and South America. Although the United States imported less than a twentieth of the total, by the 1950s the U.S.A. had a third of all New World people of African descent, suggesting that slaves were treated less brutally than in the Caribbean or Brazil. Slaveowners had distinct preferences. In South Carolina slaves from the Gambia were favored over those from Biafra as the latter were thought to be hard to control. In Virginia the preference was in the opposite direction. Genes can be used to sort out just who went where and where the ancestors of today's black Americans came from.

Many Africans have an abnormal form of the red pigment of the blood, hemoglobin. One of the amino acids (the building blocks of the molecule itself) has been changed by mutation. This "sickle cell" form protects against malaria. Although its protective role has disappeared with the control of the disease in the United States, many thousands of black Americans still carry the gene as an unwelcome record of their past. Anyone, however light their skin, who has the sickle-cell variant must have had at least one African ancestor.

The use of molecular technology on sickle cell and on normal hemoglobin allows history to be unraveled in more detail. Until recently it was known only that many people of African descent had a copy of the gene for sickle-cell hemoglobin. This revealed little more than that black Americans came from West Africa, which we knew already. Molecular biology has uncovered a huge amount of inherited variation in the DNA around the hemoglobin genes. Such diversity gives an insight into the ancestry of individual black Americans (including the great majority who do not carry a copy of sickle cell at all).

The order of DNA letters in this part of the genome has been studied in many of the peoples of Africa. The details vary from place to place. The sickle-cell mutation itself is associated with different DNA variants in Sierra Leone, Nigeria and Zaire, perhaps because it arose independently several times. Geographical variation within Africa means that the detailed structure of the gene for normal hemoglobin can also be used to track down just where the ancestors of any U.S. black originated.

Studies of American blacks show that the African patterns are matched by equivalent changes in the New World. Black Americans from the north of the U.S.A. have a different set of variants from those in the southern states. The northerners share a heritage with today's Nigerians while blacks from the south have more affinities with peoples farther west in Africa. The difference in the slave markets two hundred years ago has left evidence which remains today. Alex Haley, by comparing his genes with those from Africa, might have learned much more about his forefathers than he could hope to uncover from written records.

Many of Alex Haley's ancestors were probably not black at all. There is a specifically African form of one particular blood group, the Duffy group. Europeans have a different version of this gene. Surveys of United States blacks show that up to a quarter of their Duffy genes are of white origin (with the amount of admixture less in the southern states) probably because of interracial matings during the days of slavery. Such liaisons were covert, but widespread. Even President Thomas Jefferson is said to have had several children by his slave mistress, Sally Hemings. The biological mixing went both ways, and there are African Duffy genes among the American population which sees itself as white.

Eighteenth-century England, too, had a substantial black population. Unlike that in North America, this disappeared; not because it died out but because it was assimilated. Part of its heritage is certainly still around in the streets of modern Britain. There may be other exotic genes there as well. After all, the first slaves to cross the Atlantic were the Caribbean Indians sent to Spain by Columbus in 1495. In the sixteenth century there was a fashion for bringing members of newly discovered peoples back to Europe. The English explorer Frobisher brought back some Eskimos in 1577 and more than a thousand American Indians (including a Brazilian king) were transported to different parts of Europe. Many of these unwilling immigrants died, but some brought up families. Their legacy probably persists today.

Genes have already taken us back for hundred of years—for fifteen generations or so where black Americans are concerned. But they bring messages from far earlier in family history. Sometimes, the evidence is preserved in the corpses of our predecessors. The Egyptian pharaoh Tutankhamun was buried at about the same time as another mummy, Smenkhare. Their blood groups can still be identified. The pattern of gene sharing suggests that the two pharaohs were brothers.

Less distinguished mummies reveal even more. The dried corpse of an Egyptian child, found in the sands, contained DNA which had survived for two and a half thousand years. Amazingly enough, quite a bit was well preserved, including part of the genetic message responsible for cell-surface variation. Since then, many pieces of human fossil DNA have turned up—including some from a fifteen-thousand-year-old Australian skull. Soon it will be feasible to read ancestral genes directly and to compare them with those in the same place today. In Egypt this might test the claim of the Coptic Christians that they are the sole descendants of the ancient Egyptians, who were largely wiped out by successive waves of invasion.

It is the nature of genes that they copy themselves, so that there is no need to go back to the source to see the genes of the pharaohs or even of those with a less celebrated existence. The biology of living people gives hints about the patterns of life in times long before Tutankhamun.

For good historical reasons, more is known about the genetics of Hiroshima and Nagasaki than about anywhere else in the world. The Americans—and later, the Japanese themselves—spent many years testing whether the atom bombs had increased the mutation rate. No such effect was found, but a huge amount of information was gathered. The two cities have subtle differences in their biology. Each has a cluster of rare inherited variants not present in the other. The differences are a relic of a history which dates back for thousands of years. Hiroshima and Nagasaki were each founded by the amalgamation of different warring clans, which, eight thousand and more years ago, diverged genetically. The slight differences between the ancient tribes persist in the modern towns. Although Nagasaki was one of the few ports open to the outside world during Japan's self-imposed isolation, it had no more of an influx of foreign genes than did Hiroshima. The voices of remote ancestors echo more loudly through the two cities than do those of more recent invaders.

Such ancestral voices can even give hints about sex roles in early society. The information lies in the mitochondrial DNA. Mitochondria are small organelles in the cell which are the site of most of its energy metabolism. Each has its own piece of DNA, distinct from that in the cell nucleus. It is a closed circle containing about sixteen thousand DNA bases. Eggs are full of mitochondria but sperm have almost none. As a result, such genes are inherited only through females. Like Jewishness, they pass from mothers to daughters and sons; but only daughters pass them on to the next generation.

Mitochondrial genes evolve rapidly and are much used in the study of evolution. Their patterns of change differ from that of those which pass down through both sexes. The difference is—coincidentally but conveniently—illustrated by the evolution of first names. Boys' names do not change much from place to place, while those of girls are more localized and evolve more quickly. Only one of the top ten girls' names —Sarah—is the same in the United States and in England. Five of the ten male favorites are the same on both sides of the Atlantic: Michael, Christopher, Matthew, Daniel and David. Mitochondrial DNA evolves like girls' names: rapidly and with plenty of divergence between communities.

The mitochondria show that there has been a difference in the behavior of men and women which stretches back for thousands of years. In African pygmies, shared mitochondrial types are rarely found more than fifteen miles apart. In contrast, villages three hundred miles apart are not very different in genes which pass through males as well as females. This suggests that, at least among the pygmies, men (and their genes) traveled widely—perhaps for economic reasons (including warfare, traditionally part of the male heritage)—while women tended to stay at home.

Fossils show that humankind's earliest ancestors appeared in Africa over a hundred thousand years ago. We have African relatives too. One, the chimpanzee, has always seemed particularly close: so much so that one chimp, Flo (an inhabitant of the Gombe Stream Reserve) was the first animal to have an obituary in the London *Times*.

As any literate English-speaker knows, Tarzan of the Apes was proved to be the son of Lord Greystoke by virtue of the inky fingermarks in a childhood notebook. In spite of Edgar Rice Burroughs' clever plot, Galton had already shown that chimps themselves have fingerprints which look remarkably like those of Tarzan—or of any other human being. This suggests that chimps and men share genes. Such sharing goes much further than the fingertips. A distinguished (and famously humorless) geneticist of the 1940s once tested whether chimps share our variation in tasting the bitter chemical PTC.

He fed this stuff to three at the London Zoo. Two swallowed it with every sign of delight, but the third spat the liquid all over the famous professor. This, unscientific though it is, at least implies some common ancestry of chimpanzees and humans.

In fact, the biological affinity goes far further. One estimate, based on a test of the overall similarity of DNA, is that humans share ninety-eight percent of their genetic material with chimps. We can trace relatedness to the rest of the animal kingdom as well. It has long been known that mice and men have a lot in common. Dozens of human inherited diseases are also found in mice. Humans share even more genes with rabbits.

One optimistic evolutionist even claims to be able to find a shared structure among the genes of all living creatures, from bacteria to humans. This motif has, he suggests, persisted in some form through all the lineages which have appeared since the beginning of life three thousand million years ago. It may even represent the father (or mother) of all genes at the dawn of existence. His search for the first word of the language of the genes follows that of Pharaoh Psamtik the First, who flourished in the seventh century before Christ. Psamtik put a baby in the care of a dumb nurse and noted the sounds it made. One word was (or seemed to be) "becos," the Phrygian for bread, suggesting that perhaps the Phrygians (who then lived in what is modern Turkey) were the first people of all. Just how realistic the claim about the ur-sequence of DNA is I do not know. The scientist who published it has turned the information to a useful end. By assigning musical notes to each DNA letter he has written a sort of "symphony of life" whose theme is the gene which gave rise to everything.

Gene sharing shows the unity of life. Perhaps more important, it defines the limits of what biology can say about the human condition. Like most sciences, molecular biology usually fails to live up to its headlines. My favorite, incidentally, was one which greeted the first study of fossil DNA as "U.S. Scientists Clone Dinosaurs to Fight On After Nuclear War." Headline writers have to break a complicated story about human life into a message made up of a few letters. It is natural to think that the biological missives which everyone receives from the distant past summarize information in the same way, but they do not. A chimp may share ninety-eight percent of its genes with a typical human being but it is certainly not ninety-eight percent human: it is not human at all—it is a chimp. And does the fact that we have genes in common with a mouse, or a banana, say anything new about human nature? There have been claims that scientists will soon find the gene that makes us human. The ancestral message will then at last allow us to understand what we really are. The idea seems to me ridiculous.

Just how ridiculous it is can be seen by looking at the search for another important gene, one which I inherited from my father, and he from his and so on back to a distant ancestor that lived long before the birth of our own species. This is the gene that makes me male. The maleness gene was tracked down recently and its message spelled out in the four DNA letters, A, G, C and T. It starts like this: GAT AGA GTG AAG CGA. There are 240 of these letters altogether and, between them, they contain the whole tedious biological story of being a man. This brief ancestral bulletin does nothing to tell that half of the population which is unfortunate enough not to have it what it is really like to be male rather than female. Being a man involves a lot more than a sequence of DNA bases; and the same is true for being a human.

The Anglo-Saxon historian St. Bede—whose writings are the only real source of information about England before the eighth century—had a powerful metaphor for human existence. To him life was "As if when on a winter's night you sit feasting with your ealdormen and thegns, a single sparrow should fly swiftly into the hall, and coming in at one door instantly fly out through another. In that time in which it is indoors it is indeed not touched by the fury of the winter, but yet, this smallest space of calmness being passed almost in a flash, from winter going into winter again, it is lost to your eyes. Somewhat like this appears the life of man; but of what follows or what went before, we are utterly ignorant." Bede's allegory was a religious one but has a biological parallel. Genes have a memory of their own. By reading them there is new hope of looking beyond the hall into which our own brief existence is confined and learning something of what went before in the life of our own species; and even of guessing at what may be yet to come.

2 | The Rules of the Game

ONE OF THE MOST BAFFLING of pastimes is to watch an unfamiliar game and to try to work out what is going on. Although I lived in the United States for several years, and although the same sport is now shown on British television, I have almost no idea of how American football works. There is a clear general desire to score, but how the game stops and starts and why the spectators cheer at odd moments remains a closed book. A deep lack of interest in ball games may help in my case, but cricket is equally dull to sporting enthusiasts from other countries. They just do not understand the rules.

The rules of the game known as sexual reproduction are not obvious from its results. As a consequence, how inheritance works was a closed book until surprisingly recently. Part of the problem is that the way it does operate is so different from the way it seems that it ought to. For centuries it seemed clear that a character acquired by a parent must be passed on to the next generation. Blacksmiths' children certainly tend to be muscular and those of criminals less than honest. So obvious is the idea that it appears in the Bible. Jacob, when allowed to choose striped kids from Laban's herd of goats, put striped sticks near the parents as they mated in the hope of increasing the number he could claim. For the same reason pregnant women looked on pictures of saints and avoided people with deformities.

It took a series of painful experiments in which generations of mice were deprived of their tails to show that acquired characters were not in fact inherited. The earnest German professor who did the work would have been depressed to be reminded that Jews had been carrying out the same experiment on human anatomy for thousands of years with the same lack of success.

Another potent myth about inheritance is that somehow the characters of each parent pass to their blood, which is mixed in their offspring. Children are hence a blend of the attributes of their parents. Although this idea—a sort of genetics of the average—fails to explain why a child sometimes looks like a distant relative rather than its mother or father, it copes reasonably well with traits such as height or weight. The idea of mixing of the blood, wrong though it is, also lasted until just a few years ago. The stud book is the record kept by breeders of racehorses. A mare who had borne a foal by mating with a non-stud stallion was struck off the English stud book as her blood was deemed to be polluted and incapable of bearing a purebred horse. Indeed, a survey of elderly women in Bristol in 1973 showed that half of them believed that there was a chance of a woman having a black baby if she had sex with a black man even many years before. The horse breeders —and the crones of Bristol—had never managed to work out the instructions for the reproductive game.

Darwin himself supported the idea of inheritance by the mixing of bloods. The only section of *The Origin of Species* which does not make satisfying reading today is Chapter Six, "Laws of Variation." Darwin got it wrong in the *Origin* and, after much agonizing, developed a theory of "gemmules," in which the organs of the parents passed particles to the blood and then to sperm and egg. Offspring were, he thought, intermediate in appearance between their mother and their father.

Unfortunately, if inheritance did work in this way it would be (as Darwin later realized) fatal to the idea of evolution. The problem was pointed out by Fleeming Jenkin, the first Professor of Engineering at the University of Edinburgh. Writing in 1867—and with a sturdy disregard of today's social proprieties—Jenkin imagined "a white man wrecked on an island inhabited by negroes. Suppose him to possess the physical strength, energy and ability of a dominant white race . . . There does not follow the conclusion that after a . . . number of generations the inhabitants of the island will be white. Our shipwrecked hero would probably become king; . . . he would have a great many wives, and children . . . much superior in average intelligence to the negroes . . . but can anyone believe that the whole island will gradually acquire a white or even a yellow population? A highly favoured white cannot blanch a nation of negroes."

Jenkin was pointing out that the attributes of a distant ancestor, however valuable, can make little contribution to later generations if the ancestral bloods mix. Characters would then blend and become diluted over the years until their effects disappear. The effect is rather like putting a drop of ink into a gallon of water. However useful the ink might become at some time in the future there is no way of getting the droplet back. If inheritance worked like this, Darwin's theory would have real problems. Any advantageous character would simply be diluted out in the next generation. Fortunately, the blood myth is wrong.

It was shot down by Galton himself, who did a simple experiment. He transfused blood from a black rabbit to a white to see if the latter had black offspring. It did not. Inheritance by dilution had been disproved—but Galton had no real alternative to put in its place.

Unknown either to Darwin or to Galton the rules of genetics had already been worked out. The research was done by another biological genius of the nineteenth century, Gregor Mendel. He lived in Bohemia and published in a rather obscure scientific journal, the *Transactions of the Brunn Natural History Society*. His breakthrough was overlooked for thirty-five years after it was first published in 1865. Mendel, an Augustinian monk, attempted a science degree but failed to complete it. Like Darwin and Galton he suffered from bouts of depression which prevented him from working for months at a time. Nevertheless, he persisted with his experiments. He found that the inherited message is transmitted according to a simple set of regulations—the grammar of the genes. Later in his life (and setting a precedent for science in the present age) he was unable to continue with research because of the pressures of administration. The study of inheritance came to a halt for nearly half a century.

Grammar is always more tedious than vocabulary. Nevertheless, it cannot be avoided. The rest of this chapter explores the basic rules of genetics. Those who teach the subject still have an obsession with Mendel and his peas and I make no excuse for having them as a first course.

Mendel made a conceptual breakthrough. Instead of—like his predecessors—working on the inheritance of traits such as height or weight (which could only be measured) Mendel was more or less the first biologist to count anything. This put him on the road to his great discovery.

Peas, like many garden plants, exist in true-breeding lines within which all individuals look the same. Different lines differ in characters such as seed shape (which can be round or wrinkled) and seed color, which may be yellow or green. Peas also have the advantage that each plant carries both male and female organs. Using a paintbrush it is possible to fertilize any female flower with pollen from any male. Even a male flower from the same plant can be used. The process, a kind of botanical incest, is called self-fertilization.

Mendel added pollen (the male germ cell) from a line with yellow peas to the female part of a flower from a green pea line. In the next generation he got an intriguing result. Instead of all the offspring being intermediate, all the plants in the new generation looked like one of the parents and not the other. They all had yellow peas. This is not at all what would be expected if the "blood" of the two lines was being blended together into a yellowish-green mixture.

The next step was to self-fertilize these first generation yellow plants; their eggs were fertilized by pollen from the same individual. There was another unexpected outcome. Both the original colors, yellow and green, reappeared in the next generation. Whatever it was that produced green could still do so, even though it had spent a generation within a plant with yellow peas. This did not fit at all with the idea that the different properties of each parent were blended together. Inheritance seemed to be based on particles rather than fluids.

Mendel did more. He added up the numbers of yellow and green peas in each generation. In the first generation (the offspring of the crossed pure lines) all the plants had yellow peas. In the second, obtained by self-fertilizing the yellow plants from the first, there were always, on the average, three yellow individuals to one green. From this simple result, Mendel deduced the fundamental rule of genetics.

Pea color was, he thought, controlled by *pairs* of factors (or genes, as they became known). Each adult plant had two factors for pea color, but pollen or egg received only one. On fertilization—when pollen met egg—a new plant with two factors (or genes) was reborn. The color of the peas was determined by what genes the plant carried. In the original pure lines all individuals carried either two "yellow" or two "green" genes depending on the line they came from. When making crosses within the pure line, in each generation a new family of plants identical to their parents was produced.

When pollen from one pure line fertilized eggs from a different line new plants were produced with two different factors, one from each parent. In Mendel's experiment, these plants looked yellow although each carried a hidden set of instructions for making green peas. In other words, the effects of the yellow gene were concealing those of the green. The gene for yellow is, we say, *dominant* to that for green; which is *recessive*.

Plants with both genes make two kinds of pollen or egg, half carrying the instructions for making green peas, half for yellow. There are hence four ways in which pollen and egg can be brought together when two plants of this kind are mated. One quarter of fertilizations involve yellow with yellow, one quarter green with green; and two quarters—a half—yellow with green.

Mendel had already shown that yellow with green produces an individual with yellow peas. Yellow with yellow, of course, produces plants with yellow peas, and in a plant with two green genes the pea is green. The ratio of colors in this second generation is therefore three yellow to one green. Mendel worked backward from the ratios he had found to define this basic rule of inheritance.

Mendel made crosses using many different characters—flower color, plant height and pea shape—and found that the same ratios applied to each of them. He also looked at the inheritance of pairs of quite different characters considered together. For example, plants with yellow and smooth peas were crossed with others with green and wrinkled peas. His law applied again. There was no blending and patterns of inheritance of color were not influenced by those for shape. From this he deduced that completely separate genes (rather than alternative forms of the same one) must be involved for each character. Both for distinct forms of the same character (yellow or green color, for example) and for quite different characters (pea color and pea shape) inheritance was based on the segregation of physical units. Mendel was the first to prove that offspring are not the average of their parents and that inheritance is based on differences rather than similarities.

Biologists since Mendel have delighted in picking over his results (occasionally accusing him of fraud because some of them may fit his theories just too well). They argue about what, if anything, he thought his factors were, and speculate about why his findings were ignored. Whatever the reason for its long obscurity Mendel's work was rediscovered more or less simultaneously by several plant-breeders in the first few years of the twentieth century. His laws were quickly found to apply to hundreds of characters in both animals and plants. Mendel had the good luck, or the genius, needed to be right where all his predecessors had been wrong. No science traces its origin to a single individual more directly than does genetics, and Mendel's work is still the foundation of the whole of the enormous subject which it has become.

Mendel rescued Darwin from his dilemma (although neither of them ever knew this). A gene for green pea color or for white skin, however rare it may be, does not become diluted by the presence of many copies of genes for other colors. Instead, it can persist unchanged over the generations and may become more common should it gain an advantage.

Soon after Mendel's laws were rediscovered they were used to interpret patterns of human inheritance. Biologists cannot, of course, carry out breeding experiments on their fellow beings. Instead, they must rely on the experiments which have already been tried as humans go about their sexual business. They use family trees or pedigrees—from the French *pied de grue*, crane's foot, after a supposed resemblance of the earliest aristocratic pedigrees (which were arranged in concentric circles) to a bird's toes. Some pedigrees are fanciful, going back to Adam himself. Geneticists usually have fewer generations to play with —although one or two do trace back for hundreds of years.

The first human pedigree was published in 1905. It showed the inheritance of shortened hands and fingers in a Norwegian village. Shortened fingers ran in families and showed a clear pattern. It never skipped a generation. Anyone with short fingers had a parent, a grandparent and so on with the same thing. If an affected person married someone without the abnormality (as most did), then about half their children were affected. If any of their unaffected children married another person with normal fingers their offspring were perfectly normal and the character disappeared from that branch of the family line.

This pattern is just what we expect for a dominant gene. Only one copy (as in the case of yellow pea color) is needed to show its effects. Most sufferers, coming as they do from a marriage between a normal and an affected parent, have a single copy of the normal and a single copy of the abnormal form of the gene, one received from either parent. As a result, their own sperm—or eggs—are of two types, half carrying the normal and half the abnormal gene. When they marry, half their children carry a copy of the damaged gene. The chance of any child of a normal and an affected person having short fingers is therefore one in two. An unaffected couple never has a child showing the effects of the gene as neither of them possesses it.

Other inherited abnormalities do not behave in such a straightforward way. They are recessives. To show the effect, it is necessary to have two copies of the inherited factor, one from each parent. The parents themselves usually each have only a single copy and appear quite normal. They usually do not know that they are at risk of having an affected child. Sometimes their abnormal child looks more like a distant relative or an ancestor. Before Mendel, this was baffling. Such children were sometimes called "throwbacks" or were said to show "atavism." Now we know that they are simply obeying Mendel's laws. They have, by chance, inherited two copies of a recessive gene while their parents have just one.

The classic case of recessive inheritance is albinism. In Britain, one child in several thousand is an albino, lacking any pigment in eyes, hair or skin. Elsewhere, the condition is more common. In some North American Indians, about one person in a hundred and fifty is an albino. According to the Book of Enoch (one of the apocryphal books of the Bible), Noah himself suffered from the condition. If he did, then there is not much sign of the gene in his descendants.

Nearly all albino children are born to parents with normal skin color. These parents must each have a single copy of the albino gene matched with another copy of that for full pigmentation. Half the father's sperm carry the albino gene. Should one of these fertilize one of that half of his partner's eggs which carry the same thing, then the child will have two copies of the recessive form and will be an albino. In a marriage such as this, the chance of any child being an albino is a half times a half. This one in four chance is the same for all the children. It is not the case, as some parents think, that having had one albino child means that the next three are bound to be normal.

Patterns of inheritance in humans can, then, follow the same rules as those which apply to peas. However, biology is rarely pure and never simple. It is no surprise to learn that much of the history of human genetics has involved finding exceptions to Mendel's laws.

For example, genes do not have to be dominant or recessive. In some blood groups both may show their effects. Someone with a gene for group A and group B has AB blood, which shares the properties of both. When studying inheritance at the molecular level, the whole concept of dominance or recessivity goes away. A change in the order of bases in the DNA can be identified with no difficulty. People with two copies of the normal gene are distinct, both from those with one normal and one abnormal DNA chain and from those who have inherited two doses of the changed DNA sequence. Molecular biology makes it possible to see the behavior of genes directly, rather than having to infer what is going on, as Mendel did, from looking at the inheritance of what they produce.

Another result which would have surprised Mendel is that one gene may control many characters. For example, the sickle-cell hemoglobin variant has all kinds of side effects. People with two copies may suffer from brain damage, heart failure and skeletal abnormalities. In contrast, some characters (such as height or weight) are controlled by many genes. What is more, Mendelian ratios sometimes change because one or other of the genotypes is lethal, or advantageous.

All this (and much more) means that the study of inheritance has become a complicated subject. Nevertheless, Mendel's laws apply to humans as much as to any other creature, and they are beguilingly straightforward. Soon after they were rediscovered they were invoked to explain all conceivable—and some inconceivable—patterns of family resemblance. Long pedigrees appeared which claimed to show that outbursts of bad temper were due to a dominant gene and that there were genes for going to sea or for "drapetomania"—pathological running away among slaves. This urge for simple explanations persists today, but mainly among non-biologists. Most geneticists have had their fingers burned by simplicity once too often to believe that Mendelism explains everything.

Mendel was not concerned what his inherited particles were made of or where they might be found. He saw them just as units passed down from parents to offspring. Other scientists began to wonder just what they were. In 1909 the American geneticist Thomas Hunt Morgan, looking for a candidate for breeding experiments, hit upon the humble fruit fly. It was an inspired choice and his work with *Drosophila melanogaster* (the black-bellied honey-lover, to translate its Latin name) was the first step toward making the human gene map.

Many fruit fly traits were inherited in a simple Mendelian way. However, Morgan found some odd patterns of inheritance, which were less straightforward than those studied by Mendel. When peas were crossed it made no difference which parent carried green or yellow seeds. The results were the same whether the male was green and the female yellow, or vice versa. Some characters in fruit flies gave a different result. For certain genes—such as that changing the color of the eye from red to white—it mattered whether the mother or the father had white eyes. When white-eyed fathers were crossed with red-eyed mothers all the offspring had red eyes, but when the cross was the other way around (with white-eyed mothers and red-eyed fathers) the result was different. All the sons had white eyes and the daughters red. To Morgan's surprise, the sex of the parent which carried a particular gene had an effect on the appearance of the offspring.

Morgan knew that male and female fruit flies differ in another way. Chromosomes are paired bodies in the cell nucleus which appear as dark strands. Most of the chromosomes of the two sexes look similar, but one pair—the sex chromosomes—look different. Females have two large X chromosomes; males a single X and a smaller Y.

Morgan noticed that the pattern of inheritance of eye color followed that of the X chromosome. Males, with only one copy of the X (which comes from their mother, the father providing the Y) always looked like their mother. In females, the copy of the X chromosome from the mother was accompanied by another from the father. In a cross between white-eyed mothers and red-eyed fathers, the female offspring have one X chromosome bearing "white" and another bearing "red." Just as Mendel would have expected, they have eyes like one of the parents, in this case the one with red eyes.

The eye color gene and the X chromosome show exactly the same pattern of inheritance. Morgan suggested that this meant that the gene for eye color was actually on the X chromosome. He called this pattern "sex-linkage." Chromosomes were already good candidates for carrying genes as, like Mendel's hypothetical particles, their number is halved in sperm and egg compared to body cells.

The same patterns apply in humans. Everyone has forty-six chromosomes in each body cell. Twenty-two of these are paired, but the sex chromosomes, X and Y, are distinct. Because the Y carries very few genes, in males the ordinary rules of Mendelian dominance and recessivity do not apply. Any gene on the single X will show its effects in a male, whether or not it is recessive in females.

The inheritance of human color blindness is just like that of *Drosophila* eye color. When a color-blind man marries a normal woman none of his children is affected, but a color-blind woman whose husband has normal vision passes on the condition to all her sons but none of her daughters. Because all males with the abnormal X show its effects (while in most females the gene is hidden by one for normal vision) color blindness is commoner in boys than in girls.

Other abnormalities show the same pattern. Duchenne muscular dystrophy is a wasting disease of the muscles. Symptoms may appear in three-year-olds and patients often have to wear leg braces by the age of seven, are in a wheelchair by eleven and usually die before the age of twenty-five. Because the gene is sex-linked, it is—like color blindness—much more common among boys. Parents who have seen one of their sons die of muscular dystrophy are in the agonizing position of knowing that their other sons have a one in two chance of having inherited it.

Sex-linkage leads to some interesting differences between the sexes. For the X, women carry two copies of each gene, but men only one. This means that women contain more genetic information than do men. As we saw in the last chapter, there are two different receptors for the perception of red. Because the gene is on the X chromosome many women must carry both red receptors, each sensitive to a slightly different point in the spectrum, while males are limited to just one. Such women have a wider range of sensual experience—for color at least—than is available to any man.

Whatever the merits of seeing the world in a different way, women have a potential problem when dealing with sex-linkage. Any excess of a chromosome as large as the X is normally fatal. How, then, do females cope with two, when just one contains all the information needed to make a normal human being (or a male, at least)? The answer is surprising. In nearly every cell in a woman's body one or other of her two X chromosomes is switched off. This process was discovered by the geneticist Mary Lyon and is called, after her, Lyonization. Appropriately enough, the neatest example comes from cats. Tortoiseshell cats have a mottled appearance, which comes from small groups of yellow and black hairs mixed together. All tortoiseshells are females and come from a cross in which one parent passes on a gene for black and another transmits one for yellow hair. The coat-color gene is sex-linked. About half the skin cells of the developing kitten switch off the X carrying the black gene and the remainder that for yellow. As a result the coat is a mix of the two types of hair, the size of the patches varying from cat to cat.

The same thing happens in humans. If a woman has a color-blind son, she must herself have genes for one normal and one abnormal red receptor. When a tiny beam of red or green light is scanned across her retina her ability to tell the color of the light changes as it passes from one group of cells to the next. About half the time, she makes a perfect match but for the rest she is no better at telling red and green apart than is her color-blind son. Different X chromosomes have been switched off in each color-sensitive cell, either the normal one or that bearing the gene for color blindness.

There is another important difference in patterns of inheritance between the sexes. It involves the mitochondrial genes. When an egg is fertilized, much of its cytoplasm (including the mitochondria) is passed on to the developing embryo. Sperm pass on almost no mitochondria. As we saw in the previous chapter, mitochondrial DNA hence has its own pattern of inheritance: it is passed down the female line. It contains the history of the world's women, with almost no male interference. Queen Elizabeth the Second's mitochondrial DNA descends, not from Queen Victoria (her ancestor through the male line) but from Victoria's less eminent contemporary Anne Caroline, who died in 1881. Some genetic diseases (such as a blindness due to destruction of the optic nerve) are due to the mitochondrial DNA going wrong and are inherited down the female line. Mothers pass the gene on to sons and daughters, but only the daughters pass it to the next generation. This pattern is quite different from that of sex-linked inheritance.

These, then, are the rules of the genetical game. From here on, the rest is molecular biology: mechanics rather than physics. What genes are actually made of was established by discovering that it was possible to change the shape of bacterial colonies by inserting a "transforming principle" extracted from a relative with different-shaped colonies. The transforming principle was DNA, which had been discovered many years before in some rather disgusting experiments using pus-soaked bandages. It was the most important substance in biology.

The story of how the structure of DNA, the double helix, was established is too well known to need repeating. The molecule consists of two intertwined strands, each made up of a chain of chemical bases— adenine, guanine, cytosine and thymine (A, G, C and T for short)— together with sugars and other material. The bases pair with each other, adenine with thymine and guanine with cytosine. Each strand is a complement of the other. When they separate, each acts as a template to make its matching strand. The order of the bases along the DNA contains the information needed to produce proteins. Every protein is made up of a series of different blocks, the amino acids. The instructions to make each amino acid are encoded in a three-letter sequence of the DNA alphabet.

The inherited message contained within the DNA is passed to the cytoplasm of the cell (which is where proteins are made) through an intermediary, RNA. This ribose-nucleic acid comes in several distinct forms, each involved in transferring genetic information from DNA to proteins.

The structure of the DNA molecule is now part of our cultural inheritance. DNA is the agent of continuity between generations. The new ability to read its message—and to interfere with it—has transformed our understanding of our place in nature and our dominion over the creatures which inhabit it. It is worth remembering that the rules of the genetical game were worked out with no knowledge of where the inherited units were or what they were made of. Like Newton, Mendel had no interest in the details. He was happy with a universe of interacting and independent particles which behaved according to simple rules. These rules worked well for him, and often work just as well today.

Again like Newton, Mendel was triumphantly right; but only up to a point. Molecular biology has turned a beautiful story based on peas into a much murkier tale which looks more like pea soup. The new genetical fog is described in the next chapter.

3 | Herodotus Revised

THE GREEK TRAVELER HERODOTUS felt that he knew the world well. He voyaged around the Mediterranean and heard much of the Phoenicians' journeys into Africa. By putting what he knew of the globe's landmarks together he came to the conclusion that "Europe is as long as Africa and Asia put together, and for breadth is not, in my opinion, even to be compared with them." Herodotus had things in roughly the right places in relation to each other but the physical distances between them were hopelessly wrong.

For two thousand years maps could only be made in the Greek way. They were relative things, made by trying to fit landmarks together, with no measure of the absolute distances involved. Familiar bits of the landscape loomed much larger than they deserved. Medieval charts were not much better. Although the shape of Africa is recognizable it is greatly distorted. The cartographers' perception of remoteness was determined by how long it took to travel between two points rather than how far apart they actually were.

Genetics, like geography, is about maps; in this case the inherited map of ourselves. Not until the invention of accurate clocks and compasses two thousand years after Herodotus was it possible to measure real distances on the earth's surface. Once these had been perfected, good maps soon appeared and Herodotus was made to look somewhat foolish. Now the same thing is happening in biology. Geneticists, it appears, were until only a few years ago making the same mistakes as the ancient Greeks.

Just as in mapping the world, progress in mapping genes had to wait for technology. Now that it has arrived the shape of the biological atlas is changing very quickly. What, two decades ago, seemed a simple and reliable chart of the genome (based, as it was, on the inheritance of landmarks such as the color of peas or of inborn disease) now looks very deformed.

Morgan, with his fruit flies, discovered lots of characters which followed Mendel's laws of inheritance. Their lines of transmission down the generations were not connected to each other. There was one big exception. Certain combinations of traits, those on the sex chromosomes, did not behave in this way. Mendel had found that the inherited ratios for the color of peas were not affected by whether the peas were round or wrinkled. Morgan discovered that, quite often, pairs of characters (such as eye color and sex) traveled down the generations together. Soon, many different genes (such as those for eye color, reduced wings and forked body hairs) in fruit flies were found to share a pattern of inheritance with sex and with the X chromosome. They were, in flagrant disregard of Mendel's rules, not independent. To use Morgan's term, they were linked.

Within a few years, other traits turned out to be transmitted together, even though they were not on the sex chromosomes. Painstaking breeding experiments involving millions of flies showed that all *Drosophila* genes could be arranged into groups on the basis of whether they were independently inherited in the Mendelian fashion or not. Some combinations of characters behaved just as Mendel would have expected. For other combinations the set of traits from one parent tended to stay together in the offspring and in subsequent generations. The genes controlling them were, as Morgan put it, in the same linkage group. The number of linkage groups was soon found to be the same as the number of chromosomes. This discovery was the beginning of making the "linkage map" of *Drosophila*. Later, it became the connection between Mendelism and molecular biology.

Linkage is the tendency of groups of genes to travel together down the generations. It is not an absolute thing. Genes may be closely associated as they pass down the generations or they may show only a feeble preference for each other's company. The incompleteness of linkage is explained by some odd events when sperm and egg are formed. Every cell contains two copies of each of the chromosomes. The number is halved during a special kind of cell division in testis or ovary. During this process, the chromosomes lie together in their pairs and exchange parts of their structure. Sperm or egg cells hence contain combinations of chromosomal material which differ from those in the cells of the parents who produced them.

Perhaps, thought Morgan, this explains why within a linkage group certain genes are inherited in close consort while others have a less intimate association. He suggested that if genes are located very close to each other on a chromosome they are less likely to be parted when chromosomes exchange material during sperm or egg formation. If they are a long way apart, they separate more often. Those which are inherited independently are on different chromosomes.

Recombination, as the process is called, is like shuffling a red and a black hand of cards together. Two red cards a long way apart in the hand are more likely to find themselves split from each other when the new deck is divided than are two such cards close together. Recombination is a rearrangement of the heritage of each parent. After it has taken place, each chromosome in the next generation represents a new mixture of the genetic material made up of re-ordered pieces of the appropriate chromosome pair.

Recombination was used to make the first genetic maps. They showed the order of genes, and not much else. Like the cards in a hand held by a skilled player, genes are arranged in a sequence. Their position can be determined by how much this is disturbed each generation as the inherited cards are shuffled. Morgan suggested that two genes which are rarely separated by recombination must be close to each other on the chromosome, while those which often disengage are farther apart. By studying the inheritance of genes in twos or threes Morgan worked out their order and their relative distance apart. Combining the information from small groups of inherited characters allowed a linkage map of *Drosophila* genes to be made.

His approach—a map based on exceptions to Mendelism—was a powerful one. It has been used in bacteria, tomatoes, mice and many other creatures. Many thousands of genes have been mapped in this way. In *Drosophila* almost all have been arranged in order along the chromosomes.

Because this work requires breeding experiments, until recently the human linkage map remained—almost—a perfect and absolute blank. Most families are too small to look for deviations from Mendel's rules. As a result there seemed to be little hope of ever making a genetic map of humankind.

The one exception to this *terra incognita* was sex-linkage. Obviously, if genes are linked to the X chromosome, they must be linked to each other. It did not take long for many human traits to be mapped to the X chromosome. Even so, making the linkage map has been a painfully slow business. The gene for color blindness was mapped to the X chromosome as early as 1911. The first examples of linkage on other chromosomes did not emerge until 1955, when the gene for the ABO blood groups was found to be linked to an abnormality of the skeleton. The actual number of human chromosomes was established in the following year and the first non-sex-linked gene was mapped onto a specific chromosome in 1968. Linkage mapping in humans took a long time to get going and still has a long way to go.

Genetics has been transformed by a less ingenious but more effective approach to surveying. It does not depend on breeding experiments. Instead it involves a much more conventional kind of chart, a physical map showing the actual order of all the bases along the DNA rather than a biological map based on patterns of inheritance. The new chart has changed our views of what genes are.

The technical revolution in molecular biology showed, in an astonishingly short time, what DNA looks like and—roughly—how it works. Thirty years ago, molecular biologists were full of hubris. They had, they thought, solved all the problems of inheritance. In particular, the new ability to read the DNA would soon do the job which family studies had failed to finish. It would tell them where and in what order all our genes were. The edifice whose foundations were laid by Mendel would be complete. Optimism was, at the time, reasonable. It seemed a fair guess that the physical map of the genes would look much like a biological map based on patterns of inheritance and might in time replace it. The confidence of molecular biologists was complemented by a talent for self-publicity which would have been alien to the founders of genetics.

Optimism, at least, was quickly modified. The first explorations of the unknown territory which lay along the DNA chain showed that the physical map was very different from the linkage map. The genes themselves, far from being beads lined up on a chromosomal string, had a complicated and unexpected structure.

In the infancy of human genetics, twenty years ago, biologists had a childish view of what the world looks like. As in the mental map of an eleven-year-old (or of Herodotus) the linkage map contained only a few familiar landmarks placed in relation with each other. The tedious but objective use of a measure of distance has changed all that.

The successes of the molecular explorers depended, like those of their geographical predecessors, on new surveying instruments which made the world a bigger and more complicated place. The tools used in molecular geography deserve a mention.

The first device is *electrophoresis*, the separation of molecules in an electric field. Many biological substances, including DNA, carry an electrical charge. If they are placed between a positive and a negative terminal they move toward one or the other. A gel material (which acts as a sieve) is used to improve the separation. Gels were once made of potato starch, while modern ones are made of chemical polymers. I have tried strawberry jello, which works quite well. The gel separates molecules by size and shape. Large molecules move more slowly as they are pulled through the sieve. There are various tricks which improve the process such as reversing the current every few seconds. This means that long pieces of DNA can be electrophoresed, as they wind and unwind each time the power is turned on.

The computer on which I am writing this book can do some fairly useless things. One of its talents is—if asked—to sort all sentences by length. This sentence, with its twenty words, would line up with many otherwise unrelated sentences from the rest of the book. Electrophoresis does this with molecules. The length of each DNA piece can be measured by seeing how far it has moved into the gel. Its position is defined with ultraviolet light (which DNA absorbs), chemical stains or a radioactive label. Each piece lines up with all the others which contain the same number of DNA letters.

Another tool for surveying DNA uses enzymes extracted from bacteria. Bacteria are attacked by viruses which insert themselves into their DNA and force it to make copies of the invader. They have a defense: enzymes which cut foreign DNA. These "restriction enzymes" can be used to slice human genes into pieces. Dozens are available, each cutting a specific group of DNA letters. The length of the cut pieces which emerge depends on how often that group is repeated. If each sentence in this volume was cut whenever the word "and" appeared, there would be thousands of short fragments. If the enzyme recognized the word "but," there would be fewer, longer sections; and an enzyme cutting the much less frequent word "banana" (which does, I assure you, appear occasionally) would produce just a few fragments thousands of letters long.

The positions of the cuts (like those of the words "and," "but" and "banana") provide a set of landmarks along the DNA. Once we know where they are we have made a first step to making a physical map of the book itself based on the order of the letters and words it contains. The process is close to that carried out by the students who stormed the American Embassy in Tehran after the fall of the Shah. With extraordinary labor they pieced together secret documents which had been put through a shredding machine. By seeing how the individual fragments fitted together the students reconstituted a long, complicated and compromising message.

Molecular biology is doing the same thing. There are various tricks which allow cut pieces of DNA to be inserted into that of a bacterium or yeast. The DNA has been *cloned*. Whenever the bacterium divides, it multiplies not only its own genetic message but the foreign gene which has been inserted. As a result, millions of copies of the original piece of DNA can be made and can be studied in the exquisite detail needed for genetic geography.

Cloning was essential in making the first steps toward the physical map. It is being replaced by another contrivance, the *polymerase chain reaction*. This uses an enzyme used in the natural copying of DNA to make replicas of the molecule in the laboratory. To pursue our rather tortured literary analogy, PCR is a biological photocopier which can produce as many duplicates of a single page in the genetic manual as needed. The photocopying enzyme is taken from a bacterium which lives in hot springs, so that it is stable at high temperatures. The reaction is started with a pair of short artificial DNA sequences—primers—which bind to the natural DNA on either side of the length to be amplified. By heating and cooling the reaction mixture and feeding it with a supply of the four bases, the double strands of DNA which have been targeted unwind, copy themselves with the help of the enzyme, and re-form. Each time the cycle is repeated, the number of copies doubles and, quite quickly, millions of replicas of the original piece of DNA are generated. The inventor of PCR received a Nobel Prize in 1993.

Another piece of trickery which helps the mapmakers takes advantage of DNA's ability to bind to a matching copy of itself. DNA bases form two matching pairs; A with T and G with C. To find a gene, a complementary copy of its DNA is made in the laboratory. When the copy is added to a cell it seeks out and binds to its equivalent on the chromosome. The gene has been mapped. My computer does the same thing. On a simple command, it will search for any word I choose and highlight it in a charming purple hue. It does this best when searching for rare words (like "banana"). If a DNA probe is labeled with a fluorescent dye genes can be made to show up in the same way. The method is known as FISHing (for Fluorescent In-Situ Hybridization) for genes. A modified kind of FISH involves unwinding the DNA before it is stained. This makes the method more sensitive and gives an opportunity for yet another genetical acronym—DIRVISH (for "Direct Visual Hybridization").

All this has revolutionized the mapping of human DNA. First, it has improved the linkage map. Patterns of inheritance of short sequences of DNA can be tracked through the generations just as well as can those of color blindness or short fingers. There are millions of sites in the DNA which vary from person to person. All can be used in pedigree studies. Because most families nowadays are small, geneticists are collaborating with Mormons in the U.S.A. and with Bangladeshis in Britain, both of whom have enough children to be useful when making pedigrees. Another ingenious scheme is to use the polymerase chain reaction to multiply copies of the DNA from single sperm cells. The linkage map is made by comparing the re-ordered chromosomes in the sperm with that in the man who made them, getting over the problem of family size altogether.

Linkage mapping in humans is still a painful business. Biologists have always wanted to make a different kind of map, a map of the kind which geographers use, based not on recombination but on a straight-forward physical description of the genetic material. The ultimate map would show the order of every one of the three thousand million DNA letters. The approach is brutal: to assault the genome with time, money and tedium until the whole thing has been read from one end to the other. Already, there has been great progress. The blitzkrieg may be complete by the end of the century.

The first move in tying the linkage map to one based on molecular biology depended on a stroke of good luck. Morgan noticed that in one of his fruit fly stocks a gene which was usually sex-linked did something strange: it started behaving as if it was not on the X chromosome at all. Looking down the microscope showed what had happened. The X was stuck to one of the other chromosomes and was inherited with it. A change in the linkage relationships of the gene was due to a shift in its physical position.

Such chromosomal accidents were used in the first step toward the human physical map. Occasionally, there is a mistake in the formation of sperm or egg and part of a chromosome shifts to a new home. An accompanying change in the pattern of inheritance of a particular gene gives a clue as to where it must be.

Sometimes a tiny segment of a chromosome is absent. Those affected may suffer from several inborn diseases at once. One unfortunate American boy had a deficiency of the immune system, a form of inherited blindness, and muscular dystrophy. A minute section of his X chromosome was missing. It must have included the length of DNA which carried these genes. The American patient gave the first vital hint as to just where the gene for muscular dystrophy—one of the most frequent and most distressing of all inherited diseases—was located. The missing chromosome segment was a landmark upon which a physical map of the area around this gene could be anchored.

Gene mapping by looking for changes in chromosomes need not wait for natural biological accidents. Human cells can be cultured in the laboratory. Mouse or hamster cells can be grown, too. When mixtures of mouse and human cells are grown together, the cells may fuse, giving a hybrid with chromosomes from both species. As the hybrids divide, they lose the chromosomes (and the genes which they carry) from one species or the other. Some specifically human genes are lost each time a human chromosome is ejected. This shows where they must be. Sometimes small fragments of chromosome and the genes they carry can be associated in the same way.

The approximate chromosomal position of nearly all the common inherited diseases under simple control has now been mapped. To track them down further needs a new set of small-scale mapping systems. The techniques of mindless sequencing, as it is affectionately known, are simple. One depends on the ability of DNA to copy itself when a special enzyme is provided and the mixture fed with the A, G, C and T bases. The reaction involves "growing" copies of gradually lengthening pieces of a DNA strand from one end to the other. Four separate experiments (each using a different base) are started at the same time. Each begins the process at the same place in the DNA. By chemical trickery, some of the growing strands are stopped each time a base is added. This produces a set of DNA pieces of different length, each stopped at an A, a G, a C or a T. Electrophoresis of the mixtures on the same gel gives four parallel lines of DNA fragments arranged in increasing length. Reading across and down the gel gives the order of the bases.

This is a stunningly tedious task. One lab can only hope to carry out a minute part of the whole job, mapping a small section around the gene being searched for. These local maps—the town plans—must be put together in the right order to build up a larger map of the DNA. One way is to make a series of overlapping sequences of short pieces of DNA. This is a little like putting pages torn out of a street guide back together in the right order by looking at the overlaps at the edge of each page in an attempt to find streets which run into each other.

The method demands a certain amount of computer wizardry. There are some short cuts. One trick is to jump several pages in the street guide in the hope of missing particularly tedious parts of the neighborhood. The whole business is extremely repetitive and it has been suggested—not completely in jest—that biologists found guilty of faking their results should be sentenced to sequence a thousand, ten thousand or a hundred thousand DNA bases.

The assault on the physical map is best compared to surveying a country with a six-inch ruler, starting at one end and driving doggedly on to the opposite frontier. Sequencing will have to speed up if the task is to succeed. There still remains the brutal fact that by far the largest part of the three thousand million bases is yet to be charted. However, technology is beginning to work. Fifteen years ago, when the job began, one person could do about five thousand DNA bases a year. Now, it is routine to do twenty times as many. A set of overlapping large-scale plans has already been made for every human chromosome by inserting segments into yeast cells (a cloning technique which can cope with very long pieces of DNA). The whole genome is now covered by a linkage map in which the order and relative position of a short repeated section of DNA has been established. New methods may make the process quicker. It may soon be possible to read the order of the bases with a high-powered electron microscope scanning along the chromosome in a kind of molecular braille.

To map the whole lot will take a long time—and, some argue, is scarcely worthwhile. After all, an explorer entering a new territory does not start by making a plan of the first village and extending it in excruciating detail until the whole country is covered. Instead he picks out the major landmarks and leaves the detailed map until later, when he knows what is likely to be interesting. It may be better to concentrate on the most important sections, the genes which actually produce proteins.

There is much argument about how many of these there actually are, but one estimate has it that there may be between fifty and a hundred thousand. In most cells, most are switched off, but in the brain thirty thousand are working at any time. This is more than in any other tissue and may help to explain why more than a quarter of all inherited diseases lead to mental illness. In men, the testis come a close second to the brain in the proportion of genes which are active. When a gene is actually making something, it generates a messenger RNA (a complementary molecule which transfers genetic information from DNA to the main part of the cell). Because it makes nothing, most DNA generates no RNA at all. Extracting messenger RNAs is hence an excellent way of searching out working genes. The task is already well under way and three thousand new genes active in the brain have been found. What most of them do is quite unknown.

The complete map will probably be made in the end, simply on the grounds that it is worth doing as one never knows what might turn up. To make it will involve a reaffirmation of one of the most misunderstood facts in science: that it *is* usually possible to solve a problem by throwing money at it.

It is already clear that the physical map does not look much like the linkage maps which emerged from family studies. The central difficulty is one of scale. There are up to a hundred thousand working genes in a human being, but three thousand million DNA letters. As most genes use only the information coded in only a few thousand bases to make a protein there seems to be far more DNA than is needed. Physical mapping shows that only about five percent of it is part of a functioning gene. The human genome has an extraordinary structure and linkage maps miss most of its bizarre attributes.

We can get an idea of what the physical map looks like with a geographical analogy. Imagine the journey along the whole of your own DNA as being equivalent to one along the whole length of Britain, from Land's End to John o'Groat's via London. This is about a thousand miles altogether (which means that its American equal would be roughly equivalent to a trip from Palm Beach to New York up the Eastern Seaboard). To fit in all the DNA letters into a road map on this scale, there have to be fifty DNA bases per inch, or about three million per mile. The journey passes through twenty-three counties of different sizes. These administrative divisions, conveniently enough, are the same in number as the twenty-three chromosomes into which human DNA is packaged. The amount of DNA completely sequenced so far has covered a distance equivalent to a walk between Land's End and Falmouth, a few miles away—in the American sense, not much more than the length of the island of Manhattan. There is a long way to go yet.

The scenery for most of the trip is very tedious. Like much of modern Britain—or the northeastern states—it seems to be totally unproductive. About a third of the whole distance is covered by repeats of the same message. Fifty miles, more or less, is filled with words of five, six or more letters, repeated endlessly next to each other. Many are palindromes. They read the same backward as forward, like the obituary of Ferdinand de Lesseps—"A man, a plan, a canal: Panama!" Some of these "tandem repeats" are scattered in blocks all over the genome. The position and length of each block varies from person to person. The famous "genetic fingerprints," the unique inherited signature used in forensic work, depend on variation in the number and position of tandem repeats. To make a fingerprint an enzyme which cuts one particular repeated section is used. This slices the DNA into dozens of fragments. Other repeated sequences involve just the two letters C and A, multiplied thousands of times. Yet more of the genome is given over to occasional long and complicated messages which seem to say nothing.

It is dangerous to dismiss all this DNA as useless because we cannot understand what it says. The Chinese term "Shi" can—apparently—have seventy-three different meanings depending on how it is pronounced. It is possible to construct a sentence such as "The master is fond of licking lion spittle" just by using "Shi" again and again. This would seem like meaningless repetition to someone, like most of us, who cannot understand Chinese.

Much of the inherited landscape is littered with the corpses of abandoned genes, sometimes the same one again and again. The DNA sequences of these "pseudogenes" look rather like that of their working relatives, but they are riddled with decay and no longer produce anything. At some time in their history a crucial part of the machinery was damaged. Since then they have been gently rusting. Surprisingly enough, the same pseudogenes may turn up at several points along the journey.

After many miles of dull and repetitive DNA terrain, we begin to see places where something is being made. These are the working genes. There are some surprises in their structure, too. A functioning gene can be recognized by the order of the letters in the DNA alphabet, which start to read in words of three letters written in the genetic code, implying that it could produce a protein. Usually, there are few clues about what the protein actually does, although its structure can be deduced (and its shape inferred) from the order of the DNA letters which make it.

Many functioning genes are arranged in groups making related products. There are about a thousand of these "gene families" altogether. The best known is involved in the manufacture of the red pigment of the blood. Nearly all the DNA in the bone-marrow cells which produce the red cells of the blood is switched off. However, one small group of genes is tremendously busy. As a result they are better known than any other. Much of human molecular biology is based on this particular genetic industrial center, the beta-globin genes.

They are about halfway along the total length of the DNA; to follow the American analogy, about halfway to New York—roughly speaking at Cape Hatteras. They make some of the proteins involved in carrying oxygen. The globin industrial estate contains about half a dozen sections of DNA making related products. That for making part of the hemoglobin molecule, the red blood pigment, is quite small: about three feet long on this map's scale. A few feet away is another one which makes a globin found only in the embryo. Close to that is the rusting hulk of some equipment which stopped working years ago. The globin factory covers about a hundred feet altogether, most of which seems to be unused space between functional genes. It cooperates with another one about the same size a long way away (near Charleston on this mythical map) which produces a related protein. Joined together, the two products make the red blood pigment. Most working genes are arranged in families, either close together or scattered all over the genome.

Another surprising aspect of the map of ourselves is that genes are of very different size, from about five hundred letters long to more than two million. In nearly all the working segments are interrupted by lengths of non-coding DNA. In very large genes (such as the one which goes wrong in muscular dystrophy) the great majority of the DNA codes for nothing. The non-coding DNA participates in the first part of the production process, but this segment of the genetic message is snipped out of the messenger RNA before the protein is assembled. This seems an extraordinary way to go about things, but it is the one which evolution has come up with.

No doubt there are many more surprises waiting to turn up as the surveyors inch their way up the Eastern Seaboard. It may well be that the image of the relationship between the linkage and the physical map will change yet again. The gene geographers have already had some triumphs. The most important is a conceptual one. Because the three-letter code for each amino acid is known, it is possible to deduce the order of amino acids produced by a working piece of the DNA from its sequence of bases. What the newly discovered gene actually does can be inferred by comparing that sequence with the computer data base of others whose job is known. The process is a bit like reading a foreign language by looking up words in a dictionary to see what they mean.

There are some startling similarities between the inherited vocabulary of very different creatures. Humans have genes which are almost identical to those which control development in fruit flies, suggesting that the same basic processes are involved in both. Those making the fruit fly brain are disconcertingly close to those which produce our own. Other parallels are equally intriguing. Some genes which, when they go wrong, cause cancer are much like others which code for hormones, and, oddly enough, one human gene is almost identical to another whose only known function is to alter the pattern of veins on an insect's wing. The process of translating from one biological language to another by looking up words in the computer has already gone so far that there is a one in two chance that a newly discovered DNA sequence will turn out to be related to something else, either another human gene or one from a creature remote in the living world.

This approach has transformed human genetics. Instead of starting out with an inherited change (such as a genetic disease) and laboriously searching for its position, the opposite strategy is used—defining where a gene is and, with luck, what it does only from reading its DNA sequence. Genetics is the first science to have accelerated by going into reverse. Most human genes will be discovered using a logic precisely opposite that of Mendel: from inherited particle to function, rather than the other way around.

The first, and perhaps the greatest, breakthrough of this new approach was the tracking down of the cystic fibrosis gene in 1990. The job cost one hundred and fifty million dollars, but advances in technology mean that the cost of finding other genes will be far less. Cystic fibrosis is the most common inherited abnormality among white-skinned people. Among those of European ancestry, it affects about one child in two thousand five hundred. Its dangers have long been recognized. Swiss children sing a song which says "The child will die whose brow tastes salty when kissed." The symptoms seem at first sight unrelated, but all are due to a failure to pump salt across the membranes which surround cells. Until a few years ago children with the disease died young. Their lungs filled with mucus and became infected and they were often poorly nourished as they could not produce enough digestive enzymes. Medicine has improved the lives of those with cystic fibrosis, but few survive beyond their mid-thirties.

Family studies showed long ago that the disease is due to a recessive gene which is not on the sex chromosomes. In 1985, pedigree analysis revealed that it was linked to another DNA sequence which controls a liver enzyme, although it was not then known where that was. Within a year or so, one family showed that this pair of genes was linked to a DNA variant which had been mapped to chromosome seven. The piece of chromosome which contained the CF gene—together with many others—was inserted into a mouse cell line. It was cut into short lengths and the painful task of sequencing began. By 1988 the crucial region had been tracked down to a segment of DNA one and a half million base-pairs long. Fragments were tested to see if they had sequences in common with the DNA of other creatures such as mice. If they did, presumably the order of letters had been retained through evolution because they did something useful. Several shared sections were found. One had an order of DNA letters similar to that of other proteins involved in transport across membranes. This piece of the genome followed exactly the pattern of inheritance of cystic fibrosis. The gene had been found.

Just as in mapping the world, finding the gene was just the first step in understanding what is going on in the neighborhood. The order of the DNA letters made it possible to infer what the protein which has gone wrong looks like. Although the gene itself is a quarter of a million DNA bases long, the protein has only about one and a half thousand amino acids. Computer models of its shape show that it spans the cell membrane several times and may act as a pump.

Many families with cystic fibrosis have just one change in the protein: a single amino acid is missing about one third of the way along the molecule. This provides an immediate test for whether normal people carry a single copy of the gene without knowing it or, more important, whether a fetus might have two copies. Unfortunately, molecular mapping shows that cystic fibrosis, which once seemed a simple disorder, can be caused by more than two hundred different DNA changes. Using the genetic map to find those at risk will not be as easy as was once hoped.

Several of the most important inherited diseases have been tracked down. Huntington's Disease leads to a degeneration of the nervous system and death in early middle age. It was once called Huntington's Chorea (a word with the same root as "choreography") after the involuntary dancing movements of those afflicted. One eighteenth-century Harvard professor claimed that those with the disease were blasphemers, as their gestures were imitations of the movements of Christ on the cross. It is particularly distressing, as those at risk of having inherited the gene are left, because of the late onset of symptoms, in uncertainty about their predicament for many years. In 1983 there was a breakthrough helped by astonishing good luck on the part of the gene mappers. Almost immediately after the search started, the approximate site of the Huntington's gene was found by following its association with a linked DNA variant some distance away on the same chromosome. Then their luck ran out, and it took ten years to find the gene. It has now been tracked to the tip of chromosome four. The shape of the protein which has gone wrong—huntingtin, as it is unimaginatively called—has been worked out and, for the first time, there is hope of progress in understanding the nature of the disease.

Making the physical map is now a matter of perspiration rather than inspiration. The industrial approach is well under way. One Henry Ford among molecular biologists estimates that automation will soon multiply the rate of mapping by a hundred times and reduce its cost by the same amount. The French have taken a particularly rational approach, with their Généthon project, a factory for human genetics supported by television appeals. All this means that the physical map will probably be complete by the end of the millennium—an appropriate date.

There is a scheme to manage the process by placing it under a central organization, the Human Genome Project. Its enthusiasts see the project as a response to the biblical injunction "Know thyself." Always on the alert for a hackneyed image they refer to the sequencing crusade as the search for the Holy Grail. So far, the main product has been innocent amusement among its spectators. There is much infighting between those determined to make the American (and to a trivial extent the British) taxpayer play the reluctant role of Ferdinand and Isabella to the mappers' Columbian ambitions and those equally keen to stop them.

Opponents accuse the project of being a publicity ruse by a group who have diverted funds toward themselves from others doing more original work. The budget of the National Institutes of Health for fundamental research has dropped while that for human gene mapping has shot up. Pork-barrel biology has made an appearance as senators vie to direct work to their own state. There is a new Stalinism in science policy. Those with the politician's ear and the scientist's ego have shifted a shrinking budget into programs, institutes and centers to give the predictable new life at the expense of the unexpected. Others not in favor are squeezed out, if not into physical at least into intellectual exile. The free market in science is being abandoned in favor of the planned economy—which is odd, as those in charge of the cash are advocates of the market in every other sphere.

In their defense, the mappers point to the benefits which will emerge. Their economic value is already clear in the mad rush to patent the products of genes as they are discovered. The patents will bring fortunes to those who find treatments—or even tests—for genetic disease. Large sums are already changing hands. The rights to the polymerase chain reaction were sold to a Swiss company for three hundred million dollars. One unusually frank advocate of the Human Genome Project, while agreeing that it might make more scientific sense to survey the DNA of mice than of men, points out that the human genome is unique as it belongs to the only species which is willing to pay the bill. The cost of the scheme is much less than the race to land men on the moon—whose legacy twenty-five years later is mainly "So what?" And, unlike the moon shots, if the surveyors get only part of the way along the DNA road they will have achieved a great deal.

To look at an ancient chart—even one as faulty as that of Herodotus —is to realize that maps contain within themselves a great deal about the lives of those who drew them. They show the size and position of cities, the paths of human migration, and the record of peoples long gone. The genetic map of humanity is no exception; and it may also hold the secret to many of the diseases which afflict us. When those students now entering medical schools begin their careers—careers in which genetics will play a crucial role—a copy of the complete map should be in their hands.

4 | Change or Decay

BY THE TIME you have finished reading this chapter you will be a different person. I do not mean by this that your views about life—or even about genes—will be modified, although perhaps they may. What I have in mind is simpler. In the next half hour or so your genes, and your life, will be changed by mutation; by errors in your personal genetic message. Mutation—change—is happening all the time, within ourselves and over the generations. We are constantly being corrupted by it but, as we will see, biology provides an escape from the inevitability of genetic decline.

This chapter is about biological change, the perpetuation of error, and how progress can emerge from decay. Mutation is at the heart of human experience. It leads to old age and death but also to sex, to rebirth and to evolution. All religions share the idea that humanity is a decayed remnant of what was once perfect and that it must be returned to a higher plane by a process of salvation, of starting again from scratch. Mutation embodies within itself both decay and change: individual decline but humankind's redemption.

The first life and the first genes appeared three thousand million years ago as short strings of molecules which could make rough copies of themselves. At a reckless guess, the original molecule in life's first course, the primeval soup, has passed through three thousand million ancestors before ending up in you or me (or in a chimp or an oak tree). Every one of the billions of different genes which have appeared since then has arisen through the process of mutation. The ancestral message from the dawn of life has grown to an instruction manual containing three thousand million letters coded into DNA. Everyone has a unique edition of the manual which differs in millions of ways from that of their fellows. All this diversity comes from accumulated errors in copying the inherited message.

Like random changes to a watch some of these inherited accidents are harmful. But most have no effect and a few may even be useful. There are five thousand different inborn diseases which are due to mutation. Some show themselves when there is only one change in the DNA and some when two identical errors are passed down, one from each parent. Others manifest their effects only when there are several different changes in the genetic message. Now that medicine has, in the western world at least, almost conquered infectious disease, mutations have become more important. About one child in thirty born in the United States has an inborn error of some kind and about a third of all hospital admissions of young children involve a genetic disease. Some of the damaged genes descend from changes which happened long ago, but many others are mistakes in the sperm or egg of the parents themselves. Everyone carries single copies of mutated genes inherited from ancestors long dead which, if two copies of the same one were present, would kill them. Nobody is perfect and nearly everyone has at least one skeleton in their genetical cupboard.

Because there are so many different genes the chance of seeing a new genetic accident in one of them is small. There are a few cases when newly arising mutations can be spotted. I will concentrate on just one. Before Queen Victoria, the genetic disease hemophilia (a failure of blood clotting) had never been seen in the British royal family. Several of her descendants have suffered from it. The mutation probably took place in the august testicles of her father, Edward, Duke of Kent. The hemophilia gene is on the X chromosome; that is, it is sex-linked. To be a hemophiliac a male needs to inherit only one copy of the gene. A female needs two. The disease is hence more common among boys. This was known to the Jews three thousand years ago. A mother was allowed not to circumcise her son if his older brother had bled badly at the operation and, more remarkably, if her sister's sons had had the same problem.

As well as its obvious effects hemophilia does more subtle damage. The children bruise easily and there may be internal bleeding which can damage joints or even be fatal. Until recently, more than half the boys with the condition died before the age of five. Various treatments (including the injection of the clotting agent itself, which cures many of the symptoms of the disease) are now available and the survival rate is much higher than it was.

Several of Queen Victoria's grandsons were hemophiliacs. One of her sons, Leopold, suffered from it and the royal family history shows that two of her daughters—Beatrice and Alice—must have been carriers. The Queen herself said that "our poor family seems persecuted by this disease, the worst I know." Her Spanish descendants padded the trees in the royal park to stop their son from bruising himself. The most famous sufferer was Alexis, the son of Tsar Nicholas of Russia and Queen Alexandra, Victoria's granddaughter. Some suggest that one reason for Rasputin's malign influence on the Russian court was his ability to calm the unfortunate Alexis. The gene has disappeared from the British royal family, although there are still some hemophiliacs among the three hundred descendants of Queen Victoria alive today. Among those of western European ancestry, about one male in five thousand is affected by the disease.

Somewhat incidentally, there is a claim that another British monarch, George III, carried a different mutation. The gene causing porphyria can lead to mental illness. It may have been responsible for the well known madness of George III which led in time to his replacement by the Prince Regent. The retrospective diagnosis was made by referring to the notes of the King's physician, who noticed that the royal urine had the purple "port-wine" color which is now known to be characteristic of the disease. One of the King's less successful appointments was that of his Prime Minister, Lord North, who was largely responsible for the failure of British political will which led to the loss of the American Colonies. It is odd to reflect that both the Russian and the American Revolutions may have resulted from accidents to royal DNA.

Until about five years ago the study of human mutation was largely one of frustration, slightly ameliorated by anecdotes like these. It has been turned on its head by the astonishing advances of molecular biology. In the old days, the 1980s, the only way to study the process was to find a patient with an inherited disease and to try to work out what had gone wrong in the appropriate protein. The change in the DNA was quite unknown. This was as true for hemophilia as for any other gene. In fact, hemophilia seemed a rather simple error. Although different patients showed slightly different symptoms, the mode of inheritance was straightforward and they nearly all seemed to share the same inborn disease.

Now whole sections of DNA from normal and hemophiliac families can be compared to show exactly what has happened and, just like the genetic map itself, things have got much more complicated. Hemophilia shows how molecular biology has made geneticists' lives more difficult. First, uncontrollable bleeding is not a single disease, but several. This is because clotting itself is a complicated business. From cut to clot involves several steps. Different proteins are arranged in a cascade which responds to the damage, produces and then mobilizes the material which makes up the clot and finally assembles it into a tough protective barrier. A dozen or more different genes scattered all over the DNA cooperate in the production line.

Two are particularly likely to go wrong. One of them makes factor VIII in the clotting cascade. Errors in this gene lead to hemophilia A, which accounts for nine tenths of all cases of the disease. The other common type—hemophilia B—involves factor IX. In a rare form of the illness factor VII is involved.

Factor VIII is a protein containing 2351 amino acids. The gene is larger than most—about 186,000 DNA bases long, which, on a scale of fifty bases to the inch (the scale which makes human DNA stretch from Palm Beach to New York) means that this single gene is about a hundred yards long. For most of its length it does not produce any meaningful message at all. Only about one twentieth of its DNA codes for protein. The machinery is divided into dozens of different working sections separated by segments of apparently uninformative DNA. Much of this extraneous material consists of differing numbers of copies of the same two-letter message, a "CA repeat." Bizarrely enough, there even seems to be a "gene-within-a-gene" (which produces something quite different) in the factor VIII machinery.

The hemophilia A mutation, which once appeared to be a simple inherited change, turns out to be much more complicated than was once thought. All kinds of mistakes can happen. At least a hundred and fifty different errors have been found. Their virulence depends on what has gone wrong. Sometimes, just one important letter in the working part of the gene has changed; usually a different letter in different hemophiliacs. The bits of the machinery which join the functional pieces of the product together are particularly susceptible to accidents of this kind. In some patients part, or even the whole, of the factor VIII region has disappeared. Most remarkably, it seems that a few hemophiliacs have suffered from the insertion of an extra segment of DNA into the machinery, a segment which seems to have moved from somewhere else in the genome in the recent past.

Until recently the only way to measure the rate of new mutations to hemophilia (or to any other human gene) was to count how many sufferers there were, estimate how much damage had been done to their chances of passing on the damaged DNA and work out from this how often it must happen. Technology has changed everything. Using the polymerase chain reaction to make thousands of copies of the changed gene makes it possible to compare the genes of hemophiliac boys with those of their parents and, if they are still alive, their grandparents.

If the mother of such a boy already has the hemophilia mutation on one of her two X chromosomes, then she must herself have inherited it. The damage must have happened at some time in the past. If she has not, then her son has a new genetic accident which happened when the egg from which he developed was being formed in her own body. In a survey of all Swedish families with a son suffering from hemophilia B, several new mutations of this kind were found. However, most of the mothers of affected boys had inherited a pre-existing mutation. Surprisingly enough, most of the changed genes were not present in their own fathers (the grandfather of the patient). This means that the error in the DNA must have happened when his sperm was being formed.

A quick calculation of the number of new mutations against the size of the Swedish population gives a rate for the hemophilia B gene of about four in a million. The rate is eleven times higher in males than in females, possibly because there are more chances for things to go wrong in men (who—unlike women—produce their sex cells throughout life, rather than making a store of them at puberty). If this is true for other genes, as it may well be, then the raw material of evolution (which is what mutations are) is largely provided by males.

Most people with severe forms of hemophilia have each suffered a different genetic error. Such major mistakes probably happened in the sperm or egg which produced the child with the disorder and will disappear immediately because the child dies young. Those with milder forms of the disease often share the same change in their DNA. Their error took place long ago and has spread to many people. The presence of the shared mutation is often the first clue that these individuals share a common ancestor.

The non-functional DNA in and around the hemophilia gene is full of alterations which appear to have no effect at all and may have passed down through hundreds of generations without their carriers being aware of their presence. Near the gene itself is one of those tedious regions with many repeats of the same message. The number of copies of this repeat quite often goes up and down, so that the mutation rate in this part of the hemophilia gene is very high. The changes seem to do no damage.

Sometimes, however, this mobile DNA can be harmful. One of the better known of Victoria's contemporaries was Joseph Merrick, the Elephant Man. His skull was deformed by large bony growths which gave him his cruel nickname. To many of his contemporaries, the reason for his plight was obvious: when pregnant, his mother had been jostled by an elephant. Joseph Merrick's skeleton is still on view in the London Hospital as a mute witness to the days when inborn disease was seen as a cause of mockery. After a life of persecution and ridicule he was befriended by an eminent surgeon and found a home in the hospital, where he was visited by Victoria's own daughter. He was once thought to have suffered from neurofibromatosis, a genetic disease whose effects vary from the mild to the mutilating (although few patients show symptoms as severe as those of Merrick). Now it is believed that Merrick's disease was in fact the extraordinarily rare Proteus Syndrome. The fear by some parents who know that there is neurofibromatosis in their family that their children will be as deformed as was the Elephant Man is quite misplaced. Studies of the DNA of neurofibromatosis patients show that their disease is due to the accidental movement of one of these repeated sequences into the working part of a gene, disrupting its function.

This new fluidity of the DNA alarms geneticists as it violates the idea of gene as particle (admittedly a particle which sometimes makes mistakes) which used to be central to their lives. So powerful is the legacy of Mendel that his followers have sometimes been unwilling to accept results which do not fit into what they have been taught to believe. This is certainly true of some of the new and bizarre attributes of the process of mutation.

Scientists, in general, despise doctors. For many years, physicians reported a strange genetical effect called "anticipation." The malign effects of some inherited diseases seemed to show themselves at a younger age with each generation that passed. The effect was named seventy years ago by an enthusiastic eugenical doctor called Mott. He thought that it presaged the inevitable degeneration of society: "The law of anticipation of the insane represents . . . rotten twigs continually dropping off the tree of life." Not surprisingly, later geneticists were resistant to the idea and it disappeared from view. Now it seems that it represents a new kind of mutation, a genetic error that gets worse as generations succeed each other.

Anticipation was first noticed in a disease called the fragile X syndrome. This is the most important single cause of inborn mental defect. Many of those who show its symptoms have a constriction near one end of the X chromosome. About one male in a thousand is affected. The daughter of a mother who has fragile X is more likely to have an affected child than was her own parent although, apparently, she is passing on exactly the same gene. Mutation, of a special and surprising kind, is at work. The disease is due to the insertion of a piece of repeated DNA into a working gene on the X chromosome. Each generation, the number of copies changes: going up when it is transmitted through a female, but staying the same or decreasing when a man passes on the damaged chromosome. As a result fragile X becomes more destructive as it passes from mother to child. This is a completely unexpected kind of genetic change, which may turn out to be common.

One form of muscular dystrophy also shows more virulent effects as one generation succeeds the last. Again, a repeated DNA sequence is involved. Tracing the pedigrees of Swedish children with the disease shows that many shared an ancestor who lived in the seventeenth century. For two hundred years his descendants were perfectly normal, but suddenly some, distantly related though they now are, began to suffer from the inherited muscle weakness which is dystrophy. Studies of the area around the gene show that more copies of a DNA repeat are made each generation. Once a critical number is reached the symptoms appear. Each generation, there are more; and the effects of the damaged gene become more severe as it passes down the family line. Huntington's Disease, too, is due to an expanding DNA repeat of the three DNA letters CGG. Once more than forty copies are made, the symptoms of the disease appear; only very occasionally does anyone with less than thirty copies show any sign of nerve damage. Mistakes in the DNA are much more complicated than once seemed likely.

If the rate of mutation to hemophilia is typical, there is about one new DNA change in a functional gene per five generations in humans. This means that there are about fifty million changes in working genes per generation in the United States—which is not a small number. The actual incidence may be even higher. Studies of hormonal changes in women who are attempting to become pregnant show that eight out of ten fertilized eggs are spontaneously aborted, usually without the woman knowing anything about it. Many may carry new lethal mutations. Often all or part of a chromosome is lost during the formation of sperm or egg. The incidence of such chromosomal errors in stillborn children is ten times that among those born alive.

Why each gene has its own mutation rate and why there are so many in the first place is simply not known. The frequency varies more than a thousand times from gene to gene. Larger genes with more interspersed pieces of DNA go wrong more often than smaller ones, and certain combinations of DNA bases change more often than others. The short segments of repeated DNA outside the functional genes (such as those involved in the "genetic fingerprint") have a very high error rate. As many as one person in ten may pass on a change in this part of their DNA. The rate of mutation itself has probably evolved, too. There are enzymes which can repair the damage and when these are missing it shoots up.

There are lots of ways of increasing the number of mutations, both in body cells and in sperm or egg. In animals, radiation (such as X-rays) has a dramatic effect. On 6 August 1945 an atomic bomb was dropped on Hiroshima. Two days later, another fell on Nagasaki. They effectively ended the war with Japan—as the Emperor pointed out, in his first and somewhat understated broadcast to his people, "The situation is not necessarily developing in Japan's favor." Soon afterward, the Americans sent a team of scientists, the Atomic Bomb Casualty Commission, into the devastated cities. It set out to test whether the children of irradiated survivors of the bombs carried any new genetic damage.

A rationing system was set up which meant that all women who became pregnant could be identified and their children examined. For the first few years only birth defects and slowing of growth could be looked for. By the 1950s technology had advanced far enough to make it possible to see changes in the chromosomes of children born to irradiated parents and for twenty years after the mid-sixties there was a massive search for alterations in the structure of proteins.

The final report of the Commission (now called the Radiation Effects Research Foundation) appeared in 1991. Children were divided into two groups: those whose parents were less than one and a half miles from the burst and those farther away when the bomb fell. The task was complicated by the discovery a few years ago that the bombs had exploded some distance away from where it was once thought. Every survivor was interviewed to try to find out where they had been at the time of the explosion, whether they had been sheltered by buildings and how they had been standing in relation to the source of radiation. Not surprisingly, most people had a vivid memory of the moment of the bombing and from this it was possible to estimate the dose which each received.

More than a million genes coding for proteins were looked at. Thousands of electrophoretic gels were tested to see if there were any shifts in the positions of bands in children compared to those in their parents. Three mutations were found in children whose parents were in the cities when the bombs fell and three in the children of those outside the fallout zone. These people have probably had more radiation than any others in history. The actual amount which each received is not certain as new work on chemical changes produced by radiation in the concrete which survives from 1945 suggests that the dose was even higher than estimated at the time. Nevertheless, the rate of DNA damage does not seem to have increased much.

Mutation at these protein-coding genes is such a rare event that even a doubling of the rate in the exposed population might have been missed in the Atomic Bomb Casualty Commission survey, enormous though it was. A study of a quarter of a million births in Hungary after the Chernobyl nuclear accident also gave no sign of an increase in the numbers of children born with genetic disease.

However, there is no doubt that radiation can cause inherited mutations in humans. Up to two thirds of the sperm cells of cancer patients who have been given large X-ray doses carry chromosomal changes. There is also enough evidence from other animals to make lower doses of radiation a real cause for concern, particularly as there is a clear link between agents capable of causing mutations in sperm and egg and those which lead to cancer (see p. 81). The biggest avoidable source is radon gas, which leaks from granite. People living in granite houses in Cornwall may be exposed to more excess radiation than are those working in nuclear power stations. In the United States, houses were built using radioactive sands in the foundations. Their occupants faced twenty times the average radiation dose, and their houses are being demolished. Even those at lower risk are advised to install fans to stop the gas accumulating. There are other sources of radiation (such as flying, medical X-rays and even luminous watches) but for most people these involve very small doses.

Chemicals are probably more important. The number of chromosome errors in nuclear power-station workers is slightly greater than that of the general public: but the number in those working in coal-fired stations is even higher because of the noxious chemicals which are by-products of burning coal. A breakthrough in this work was to use bacteria to test whether particular chemicals were harmful. A huge number of likely, and some unlikely, possible agents were tested. Some, such as those once used in hair dyes, were found to have a powerful effect and have been banned. Others, including those in black pepper, in Earl Grey tea and in some pesticides, also caused mutations. Some of the most potent are perfectly natural. This is not surprising, as plants produce many toxic chemicals for defense against insects. Even lettuce, the epitome of healthy eating, contains chemicals which cause mutations in mice.

Cynics argue that organic foods are more dangerous than food which has been sprayed because of the noxious chemicals found in the molds which grow on them. Fortunately, fresh fruits and vegetables seem to reduce the rate of mutation. Even an increase in temperature can increase the mutation rate and to cool the nether regions by wearing the kilt in the granite city of Aberdeen should help to reverse any effects of radon gas.

Mutations, however they arise, are the raw material of evolution. Humans—and all other living creatures—change over the ages by accumulating them, but make this change without degenerating. Life progresses; it does not decay. However, every individual is mortal. As they grow older the bodily machinery corrodes until it finally breaks down.

Part of this erosion comes from genetic changes within our own bodies. To build a human being from a fertilized egg involves making hundreds of millions of cells, each with its own copy of the original genetic message. As the copying process is imperfect there are plenty of chances for mistakes. Even in adult life most cells continue to divide. Red blood cells are renewed every four months or so, for example. Millions of cells divide each second and each minute every individual produces thousands of miles of newly copied DNA. As a result, everyone accumulates huge numbers of new mutations in body cells during their lives. Each individual is an evolving system whose identity changes from day to day.

Some of these changes can lead to disaster. Many cancers result from genetic accidents similar to those which produce inborn abnormalities. In the past decade or so some cancers have begun to look increasingly like genetic diseases. They arise either as simple errors in the DNA of body cells, or as an inherited predisposition to cancer which is sparked off by something in the environment. There are a hundred or so specialized genes which control the growth of cells. When they mutate, cells may multiply uncontrollably. Just as in hemophilia, all kinds of mistakes can happen. A single DNA base may change or whole sections of the message be lost. Sometimes the error involves genes moving from one chromosome to another and occasionally—and amazingly —viruses carry cancer genes similar to those of humans and insert them into the DNA. Often, several different genetic accidents are needed to promote the development of cancer. The general picture is not very different from that of mutation in sperm or egg.

Just as for inherited mutations in the germ cells, radiation and chemicals increase the chance of damage. For some genetically unlucky people even a small amount of sunlight may cause cancer of the skin. A few have a faulty gene which means that they cannot repair damage to DNA, but for many more the risk comes from their combination of genes for red hair and pale skin which allows more ultraviolet into the cells. The dials of luminous watches in the First World War were painted with a radioactive material. Many of the unfortunate women who did the work had the habit of licking their fine brushes. Most died of a cancer called "phossy jaw." Many causes of lung cancer arise from exposure to radon. There seem to be some clusters of childhood leukemia around a few nuclear power stations, but these are puzzling as the amount of radiation emitted seems far too low to cause the effect directly. For most people, exposure to radiation is so low that it cannot be an important general cause of cancer.

Once again, chemicals play a part. Those in tobacco smoke are potent agents of genetic damage when tested on bacteria. Some industrial chemicals are just as bad. Alcohol, too, is far from blameless, particularly when, as so often, a drink is accompanied by a cigarette. Certain chemicals bind to DNA to cause their damage. Measuring the amount of bound chemicals gives an estimate of exposure to mutagens. In some places the results are alarming. The Polish city of Gliwice is one of the most polluted places in the world. Much of the pollution comes from burning soft coal. Gliwice has a high rate of cancer. Many of its inhabitants have large amounts of poisonous chemicals bound on to their DNA. The amount goes up sharply in the winter, when pollution is at its worst. Many of those exposed are likely to develop cancer.

Cancer represents a decay of the genetic message and a loss of control of cells by the DNA they contain. Age may reflect the same process. As our bodies are in a constant fever of DNA replication, the older we are the more divisions there have been and the more chance for mistakes. The cells of a newborn baby are separated by only a few hundred divisions from the egg; but mine, at the age of forty-nine, are distanced from it by thousands. My genes have had more chances to mutate than have those of a baby. What is worse, they are less effective at repairing the damage. The effects of mutation in older people can be seen directly. Their cells may contain altered genes which are producing inappropriate proteins. For example, many elderly Europeans have a small but noticeable amount of sickle-cell hemoglobin in their blood. This gene is normally found only in Africans but has, in their case, appeared as a new mutation within their aging bodies.

All this helps to explain why cancer is largely a disease of the old. The biological identity crisis which we define as old age and which is solved by death happens when the genetic message has become so degenerate that its instructions no longer make sense. The rate of aging is programmed. Mouse cells in culture stop dividing after about four years, while human cells can carry on for nearly a century.

Just like making a series of photocopies, one from the other, a little information is lost whenever a cell divides. Parts of the message disappear with time. DNA is packaged into chromosomes. Each has a specialized length of DNA at its end. This gets shorter with age. In a baby it is about twenty thousand letters long, while in a sixty-year-old it is less than half this. Cells from tumors have lost even more DNA from the chromosome ends. About four letters are dropped from this section of the message each time a cell divides, so that an old body is working from an imperfect instruction manual, full of printing errors. Just the same thing happens to mitochondrial genes, which are shot full of errors as age continues its inexorable progress.

Age may itself be due to the accumulation of genetic accidents. Human cells in culture age more quickly when they carry a defect which increases the mutation rate and some children who inherit a tendency toward cancer also show the symptoms of senility more rapidly. The immune system, which has the highest mutation rate of all body cells, is among the first to fail with age. It seems that the decay of our elderly selves is, at least in part, due to mutation.

Age also increases the number of genetic errors in sperm and egg. In the Swedish hemophilia survey, fathers whose daughters carried new mutations were eight years older than the average Swedish father. It is more than an accident that Victoria's father was himself more than fifty when she was born. An old father's sperm are separated by many more generations of cell division from the fertilized egg which produced him than are those of a younger male. The effects of age can be striking. The chromosomal mutation which gives rise to Down Syndrome is thirty times more common among mothers older than forty-five than in teenagers. The rate of spontaneous abortion, too, goes up by five times between the ages of thirty-five and forty-five, perhaps because of the accumulation of chromosomal damage.

All this adds a certain irony to the claims of one institution devoted to reversing the decay of the human race, the Center for Germinal Choice in California, in which Nobel Prize–winners make genetic deposits for hopeful mothers. The depositors may once, as they claim, have approached genetic perfection, but because they are elderly that perfection has been marred by age.

Why, if our genes change and decay through our lives, does the human race not degenerate as one generation succeeds another? The answer lies, it seems, in sex. To define sex is simple; it is a way of enabling genes from different ancestors to be brought together in the same person. As the next chapter shows, sex provides a chance to purge ourselves of the harmful mutations which arise each generation. Sex is, in more ways than one, the antithesis of age.

Nearly every novel, play or work of art revolves around the eternal triangle of sex, age and death. All three—and our very existence—are aspects of the same thing; of errors in the transmission of genes, of mutation. Humanity is not a decayed remnant of a noble ancestor. Rather we are the products of evolution, a set of successful mistakes. Genetics has provided the solution to one of our oldest questions; why people decay, but humanity does not. In one sense at least, salvation lies in the genes.

5 | The Battle of the Sexes

BIOLOGISTS have an adolescent fascination with sex. Like teen-agers, they are embarrassed by the subject because of their ignorance. What sex is, why it evolved and how it works are the biggest unsolved problems in biology. Sex must be important as it is so expensive. If some creatures can manage with just females, so that every individual produces copies of herself, why do so many bother with males? A female who gave them up might be able to produce twice as many daughters as before; and they would carry all her genes. Instead, a sexual female wastes time, first in finding a mate and then in producing sons who carry only half of her inheritance. We are still not certain why males exist; and why, if we must have them at all, nature needs so many. Surely, one or two would be enough to impregnate all the females but, with few exceptions, the ratio of males to females remains stubbornly equal throughout the living world.

The obsession with sex is an ancient one. The earliest art works are overtly sexual. The Dancing Venus of Galgenberg, an elegant serpentine statuette without the exaggerated breasts and buttocks found in later variations on the theme, is around thirty thousand years old. An interest in the female form goes back even further. A small pebble from an excavation in Israel has been grooved to resemble a woman's body. It may be as much as eight hundred thousand years old, making it the oldest known work of art.

Curiosity about the meaning of sex is also nothing new. Plato, in the *Symposium,* suggested that there were once three sexes; males, females and androgynes or hermaphrodites. The third sex was split apart by an angry Zeus and doomed to spend eternity forever seeking its partner: as Plato puts it, "Zeus moved their privates to the front and made them propagate upon themselves. If, in all these claspings, a man should chance upon a woman, conception would take place and the race would be continued; whereas if man should conjugate with man, he might at least obtain such satisfaction as would allow him to turn his energies to the everyday affairs of life." This provided Plato not only with an explanation for the origin of sex and the sex ratio but a neat way of explaining the variety of sexual attractions common from ancient Greece to the present day. Two thousand years later the English wit Sydney Smith had the same idea, although his three sexes were men, women and clergymen.

To define sex is easy enough. It is a way of producing individuals who contain genes from more than one line of descent, so that inherited information from different ancestors is brought together each generation. In an asexual creature everyone has one mother, one grandmother, one great-grandmother and so on in an unbroken chain of direct descent from the ur-mother who began the lineage. Sexual organisms are different: the number of ancestors doubles each generation. Everyone has two parents, four grandparents and so on. When sperm or egg are formed, each has half the number of genes present in body cells and in each the genes are scrambled into new arrangements by recombination (see p. 57). After mating, the novel arrays come together to produce a new and unique individual. Reshuffling the genetic message is at the heart of sexual reproduction.

The meaning of sex is illustrated by two eponymous heroes of British history, King Edward VII (who flourished in the years before the First World War) and the King Edward variety of potato (which has fed the British working class for almost as long). The potato, unlike the royal family, reproduces asexually. Every King Edward potato is identical to every other and each one has the same set of genes as the hoary ancestor of all potatoes bearing that name. This is convenient for the farmer and the grocer, which is why sex is not encouraged among potatoes. King Edward himself was a very different kettle of fish. Half his genes came from his mother, Queen Victoria, and half from his father, Prince Albert. He himself was a new and unique genetic mixture who combined some of the qualities of the two and of an ever-widening pool of more distant ancestors.

That is what sex is. To understand why it exists is more difficult.

One theory about why life is not female has to do with mutation. If a sexless organism has a harmful change to the DNA it will be carried by all her descendants. None of them can ever get rid of it, however destructive it might be, unless it is reversed by another change in the same gene—which is unlikely to happen. In time, another damaging error will occur in a different gene in the family line. A decay of the genetic message will set in as one generation succeeds another, just like the decay that takes place within our aging bodies as our cells divide without benefit of sex. In a sexual creature the new mutation can be purged as it passes to some descendants but not others. Sex also has a more positive effect on evolution: as the environment changes (as it often does) some of the new combinations of genes are particularly able to cope with the new challenges.

Very few animals have given up sex. They include the odd lizard or fish, but none of our close relatives. Even creatures like aphids which manage without it for most of the time require a bout of sexual reproduction once a year or so. With occasional exceptions such as rotifers (tiny creatures living in fresh water, among whom no male has ever been found), asexual lineages all seem to come from recent ancestors with a normal sex life, suggesting that chastity is an evolutionary dead end. Just why abstinence is a bad thing is not certain. In spite of the attractions of the mutation theory the frank answer is that, although the reason for the existence of women is obvious enough, nobody has any real idea what point there is in being a man.

Men have, however, made many attempts to justify their existence. They point out that creatures which have given up males do have problems. Nearly all asexual plants can only be used for a few years. They become so loaded with genetic damage that they no longer thrive, or they cannot keep up in the evolutionary race with their parasites who in time prevail. Their lineage has become senescent.

Potatoes are a good example of the risks of celibacy. The Irish potato famine happened because the potatoes used nearly all belonged to an old and sexless variety. In the mid-nineteenth century every European potato was descended from one or two introductions from the New World made three hundred years earlier. The new crop quickly spread throughout Europe. Louis XVI of France, in a cunning exploitation of the rustic mind, put guards on the first potato fields during the day but quietly removed them at night. The peasants, impressed by the apparent value of the crop, were quick to steal examples and to grow them in their own fields. In Ireland in 1840 every adult ate around ten pounds of potatoes a day (largely because their grain was exported to England to pay rent to their expatriate landlords). Famine struck with amazing speed and devastating effect. In 1845, the Irish *Freeman's Journal* wrote "We regret to have to state that we have had communications from more than one correspondent announcing the fact of what is called 'cholera' in potatoes in Ireland, especially in the North. In one instance the party had been digging potatoes—the finest he had ever seen—from a particular field, and a particular ridge in that field until Monday last; and digging in the same ridge on Tuesday he found the tubers all blasted and unfit for the use of man or beast." In the next five years, one and a half million Irish people died from starvation. Their crop had been attacked by a fungus, the potato blight, which is usually sexual and has many generations to each one among its hosts. The parasites evolved more quickly than could the potato. Nowadays, plants with new sets of genes are tried every few years to stop this happening. Other asexual crops, such as bananas, have as yet escaped the fate of the Irish potato (although it cannot be indefinitely delayed). The potatoes were forced into an evolutionary dead end from which the only escape is sex.

The perils of abstinence can be seen in one chromosome which has abandoned sex, in the strict genetical sense at least. The Y chromosome is present only in males. When germ cells are formed, all the other chromosomes line up next to each other—chromosome 21 with chromosome 21, or X with X, for example—and indulge in recombination, the orgy of exchange of genetic material described in Chapter Three. In a male, the Y does line up with the X, but its embrace of its fellow is less than enthusiastic. Only the tip of the Y exchanges genetic material with the X. The rest of the chromosome is held in a kind of genetical purdah, safe from the advances of other genes.

Abandoning sex has had terrible effects on the Y chromosome. It has lost nearly all its functions apart from those few involved in determining masculinity. Instead there are long sequences of apparently meaningless DNA letters, many of which are repeated thousands of times. Perhaps this is a hint of what might happen to asexual lineages if abstinence goes on for long enough. Mutations accumulate and cannot be shed and junk DNA may creep in and prove impossible to dislodge. Apart from its narrow role in ensuring the persistence of men, the Y chromosome embodies within itself an awful warning of the dangers of continence.

Sex means that new mixtures of genes arise all the time as the chromosomes from each parent recombine. Each generation produces successful individuals who have been dealt a favorable hand of mutations while others inherit a less advantageous set and fail to pass them on. George Bernard Shaw illustrated this in a hackneyed but biologically accurate phrase. When an actress asked if she could bear his child, who might have her body and his brains, Shaw pointed out the risk of producing an infant with her brains and his body. Sex reshuffles life's cards: it produces beautiful geniuses who survive and ugly fools who do not.

It controls the fate of the thousands of new genetic errors which appear each generation. Some are harmful, some not. Sex is a convenient way of bringing together the best (some of which may even be better than what went before) and purging the worst. It separates the fate of genes from that of those who carry them. Sex is a kind of redemption, which, each generation, reverses biological decay. In some ways, sex is the key to immortality. It is the fountain of eternal youth—not for the individuals who indulge in it, but for their genes. Sex speeds up evolution because each generation consists of new and unique mixtures of genes, rather than thousands of copies of the same one. Instead of always drawing the same hand in life's card game (which might be successful in one encounter but which is unlikely to be so in all), every fertilized egg has a new deal and a new chance to win in the struggle for existence. The chance may be a small one, but as so many hands are dealt sex becomes a worthwhile, albeit expensive, way of gambling against a hostile world.

Sex is universal, fascinating and complicated. One of the surprising discoveries made when comparing the physical map of the DNA with the linkage map (which is, as we saw in Chapter Three, based on recombination) is that some parts of our DNA are sexier than others, at least in the sense that more recombination takes place at these appropriately named "hot spots."

There is another baffling and subtle problem—if sex, why sexes? If recombination, the shuffling together of the genetic material of two individuals, is such a good thing why has evolution not come up with a scheme which allows everyone to mate with everyone else? As we are limited in our choice of partners to those of a different sex, having just two sexes seems to be very inefficient. Nearly all organisms (with the exception of a few single-celled creatures which have up to six sexes) exist as just males and females. This means that only half the population is available as a potential mate. If there were three sexes, then two thirds of the group might be accessible, and a hundred different sexes could make ninety-nine percent of our fellows into possible partners. One answer (and it is only one of several) has to do with what seems at first sight the antithesis of sex—conflict.

Males are best defined as the sex with small sex cells, sperm; and females as that with large, eggs. Body cells contain not only the DNA in the nucleus, but a lot more in the cytoplasm which surrounds it. Some is associated with mitochondria (which have genes of their own: see p. 39). In many creatures, there is even more DNA in the cytoplasm. It comes from what may once have been simple free-living creatures which have taken to hitching a ride in their cells. This DNA (like that in the nucleus) has its own agenda, which is to be copied and passed to the next generation. The cytoplasm is its territory; and just like a blackbird or a tiger it defends its homeland against invaders. If sperm and egg were the same size (and each had its own population of extraneous DNA) there is a danger of war breaking out on fertilization. Then, two sets of cytoplasmic genes suddenly occupy the same territory in the fertilized egg. Just like tigers (and like those few simple plants which have sexes with equally sized sex cells) one set might attack the other until it prevails. This is expensive and time-wasting and may even harm the genes in the nucleus.

The dispute is resolved by one sex—the males—unilaterally giving up the struggle. The sex which surrenders passes on none of its cytoplasmic genes (which are excluded from the tiny sperm) while the winner, which makes the egg, passes on large numbers in its mass of cytoplasm. As in most wars, the only stable number of opponents is two: and the existence of males and females (rather than dozens of different genders) represents a truce in the battle of the sexes.

Although biology has only a vague idea of why sex is there and why it is limited to the tedious dualism of male and female, it is beginning to find out a lot about how it works. The technical revolution in genetics has shown what a simple thing sex is in—and at—conception and what a complicated tangle it becomes later in life.

Existence is, it seems, essentially female and masculinity just a modification of the feminine experience. The gene which produces maleness is simple; so simple indeed that it has given impetus to a new (and to some of us rather depressing) theory of the origin of sex, that masculinity may originally have arisen as a piece of selfish DNA. This theory suggests that males are just parasites on females, individuals who have the pleasures of reproduction (and of passing on their maleness gene) with few of the pains.

The Y chromosome forces a developing embryo into masculinity. If, for some reason, the Y is missing then the fetus develops as a female. Some children are born with an extra X chromosome. Their chromosome set is XXY. Such individuals are male (although they are sterile). People with several X chromosomes and a Y have been found and these too are male, emphasizing the power of this small chromosome in imposing its function on the X.

The actual gene which determines human sex was tracked down following the discovery of a very few males with two X chromosomes. This is in apparent contradiction to the rule that to be a male needs a Y. In these men (most of whom know nothing of their condition) a tiny part of the Y chromosome has been broken off and attached to an X. This X is then armed with the information needed to inflict maleness. Because the segment of transferred Y chromosome is so small the augmented X was very useful in tracking down the crucial gene, which is only 240 DNA bases long. It is found in all male mammals and is similar to a gene which determines what passes for masculinity in yeast.

Although the machinery for deciding the sex of a fertilized egg is simple, the road to adult gender is a complicated and difficult one. Sexuality is a flexible thing. In some creatures, this is obvious. In crocodiles, sex is determined by the temperature at which the eggs develop, so that females must lay their clutches in a place with a temperature range which allows both males and females to be produced. In certain fish, embarrassment—or social pressure—is important. A shoal of females is guarded by a male. If he is removed there is a period of confusion, until one of the females changes sex and assumes his role.

Sex determination is hence less rigidly programmed than might at first appear. The nature of the switch from female to male varies from species to species. Even in those in which it is determined early in development there are many chances to take one turning rather than the other on the road to adulthood. The male-determining gene switches on a cascade of different hormones. Sometimes, these go wrong, and there is a whole range of hermaphrodites and intersexes which are due to failures in one step or the other in the sexual chain.

Once sexuality gets started enormous consequences flow from it. Most of natural history is the scientific study of sex, as the characters which differentiate birds, insects and flowers from each other are largely associated with reproduction. The diversity of sexual choices in the living world means that comparing the sex lives of different creatures may say a lot about how sex evolved and why animals behave the way they do. Although humans are in many ways distinct from the rest of creation, it might be possible to learn something about our own reproductive habits by looking at those of other species.

Many people have attempted to draw sweeping conclusions about humankind from studies of the private lives of monkeys and apes. It is always dangerous, and usually futile, to try to explain human behavior in the simple terms which apply to animals. Attempts to do so usually fall into the "pathetic fallacy," the literary trap which sees emotions mirrored in the weather or the landscape. Occasionally—very occasionally, as in *Wuthering Heights*—this works, but it usually ends in bathos. Anthropology has the same problem. It is fatally easy to read into the animal world what we would like to see in our own, to explain the human condition as an inevitable consequence of biology. Even Charles Darwin, a veritable Brontë among sociobiologists, was at fault. Hidden in his unpublished notebooks is the damning phrase "Origin of Man now proved—metaphysics must flourish—he who understands baboons will do more towards metaphysics than Locke."

Metaphysics is one thing, sex another. The Nobel Prize–winning animal behaviorist Konrad Lorenz saw humans as "killer apes" anxious to pass on our own genes by murdering the opposition, which may have explained his own early flirtation with the Nazis; and any decent airport has a row of paperbacks whose embossed covers purport to explain human nature as emerging from a history as primates with one or other sexual and social preference. Until recently the study of sexual behavior was little more than a set of unconnected anecdotes. It has been transformed by the rebirth of one of the oldest techniques in biology. Comparative anatomy is what convinced Darwin that men and women are related to monkeys and apes. Now there is a new science of comparative behavior which reveals a great deal about how and why sexual conduct evolved.

As many people know to their cost, sex is filled with strife. The very existence of males and females is the resolution of a war to pass on cytoplasmic genes. There is also conflict between males to find a mate and between males and females as they invest time and effort in bringing up the young. Sometimes the disputes are obvious. There is a struggle between males, leading to the evolution of spectacular organs such as red deer's antlers, which are used by the winners to monopolize the females. Other characters—such as a baboon's brightly colored face —are more subtle statements of male talent and may evolve because they are preferred by the opposite sex.

There is little evidence (in spite of much prurient speculation about beards, breasts and buttocks) that humans have attributes of this kind but, as in most animals, conflict between human males is greater than between females. Being a man is a dangerous thing. At birth there are about 105 males to every 100 females, but this drops to 103 to 100 at the age of sixteen and by the age of seventy there are twice as many women as men. Men have more accidents, more infectious diseases and kill each other more often than do women. The murder rate, an almost exclusively male preserve, peaks at the age of twenty-five both in London and in Detroit (although the actual rate in the latter is forty times higher than in London). This is close to the prime age of reproduction. Oddly enough, eunuchs and monks live for longer than do males condemned to a normal sex life.

Our close relatives have very different life-styles. From the human perspective chimps are deplorable but gorillas dull. A male chimpanzee copulates hundreds of times with dozens of females each year. The faithful gorilla, on the other hand, has to wait for up to four years for his female to be ready to mate after she has given birth, and even then she is available for only a couple of days each month. Not surprisingly, there is intense competition among gorilla males for access to females. A successful male may accumulate half a dozen of them, which leaves, of course, a number of gorilla wallflowers out in the cold and anxious to fight for their reproductive rights. Often, these fights are savage—not surprisingly, since what is at stake is the male's evolutionary future. Humans are unusual. They live in cooperating groups as (more or less) faithful pairs. In this they are more similar to seagulls than to most apes. The closest in behavior to ourselves is the pygmy chimpanzee. This is less studied than its larger relative, but seems to form long-lasting pairs within a stable group of individuals and has other attributes not unlike our own (such as face-to-face copulation). The average Frenchman or Briton has ten sexual partners during his life. As in many primates, there is more variation among men in their success in finding partners than among women. One percent of the men are responsible for sixteen percent of the total number of female partners.

Among the primates there is a good fit between the size difference of males and females and patterns of mating. In those species with large harems and angry bachelors, males are much bigger than females, presumably because bulk and aggression are helpful when battling for a mate. Gorilla males are twice as big as females, while the chimpanzee's more relaxed life-style has taken the pressure off sexual hostility and males and females weigh about the same. The argument from anatomy suggests that humans, with men just a little larger than women, have a history of mild polygamy intermediate between that of chimp and gorilla.

The human mating system is, of course, flexible and can shift quickly (as in the recent change toward serial monogamy—constancy within a relationship, but more than one relationship in a lifetime). There seem to be some general rules. Strict monogamy is rare. In most societies, men have more than one mate during their lives. Polygamy (one male with several wives at once) is far more common than polyandry, the opposite pattern, although this exists in Tibet. In polygamous societies, as a few men have many wives some must have none.

The primates illuminate other skirmishes in the battle of the sexes. Some suggest a more salacious past for humankind than does the modest difference in the size of men and women. The struggle between males does not stop once mating has taken place. There is competition between sperm, too. Often a female will use the sperm of the male she mated with last, which means that a successful sperm donor must ensure that no other male mates with her until the eggs are fertilized. This is why dogs stay paired after copulation. The male is guarding the female against intruders.

An even less subtle way of ensuring the success of one's own sperm is to flood out the contribution of the preceding male. There is, among different species of primate, quite a good fit between the size of the testes and the extent of male promiscuity. Chimpanzees, the Lotharios of the primate world, have enormous testes while gorillas, in spite of rumor to the contrary, are far less well endowed. Humans, surprisingly enough, are not too different from chimps in this respect—which may say some startling things about our past. Real enthusiasts for evolutionary explanations point out that men produce more sperm when returning to their partner after a long absence, possibly to overwhelm any alien sperm which may have intruded. There is also the question—as yet unanswered by science—as to why, in penis size, man stands alone. There are limits to what biology can explain and this may be beyond them, although it does seem that there is a new genre of evolutionary pornography just waiting to be written.

Charles Darwin was not the first to see that differences in sexual success might be important in evolution. James Boswell in his *London Journal* (which reveals him to have been no mean sexual performer in his own right) wrote—rather piously, given his own behavior—that "If venereal delight and the power of propagating the species were permitted only to the virtuous, it would make the world very good." Darwin noticed, however, that sexual selection (as he called it) might do much more than improve a male's ability to defeat his ardent competitors. He was much concerned with the evolution of characters with no obvious biological advantage (such as the peacock's tail—or the large human penis, for that matter). The struggle for sex might, Darwin thought, have more subtle consequences than just the evolution of large and aggressive males. If females prefer, for one reason or another, a particular male attribute (such as a brightly colored tail), then males who have it will reproduce more successfully. The bright tail or its equivalent will become more common in later generations and the showiest males will once again be preferred by the females. In time there may evolve bizarre structures which are so expensive to the unfortunate males who carry them that they can evolve no further. Female choice may, Darwin suggested, be as important a part of the sexual equation as is male aggression.

In his book *The Descent of Man and Selection in Relation to Sex*, he went further. He suggested that mating preferences explained why human races looked so different. It was not that they had evolved to fit the place in which they live, but as a consequence of arbitrary choice of a mate. In different places, those looking for a partner may have made different—and quite capricious—choices. In time, the people of the world diverged: for example, Darwin speculated, those with darker skins might have been seen as more attractive in Africa and those with lighter in Europe. There is certainly plenty of evidence that people tend to marry others who are similar to themselves in intelligence, color and—strongest of all—length of the middle finger, but there is as yet no real evidence that such choices are important in evolution.

Men do tend to agree in their estimation of how attractive a particular female might be. Galton himself had the idea of making composite photographs, in which the pictures of a number of society beauties were printed one on top of the other in the hope of producing something close to the ideal woman. His Ms. Averages look rather insipid to the modern eye. The same thing can now be done by computer. For both male and female faces most people find an image made up of several individuals more attractive than one based on a single person: and the more faces used in making the computer image the more appealing it is judged. Why there should be this triumph of the typical is not certain (although there have been wild and unsubstantiated conjectures that those with extreme faces might also have aberrant—and less desirable—genes).

Any discussion of the evolution of sex seems doomed to stray for a time onto such untamed shores of speculation. One theory of why males may carry eccentric characters is known as the handicap principle. It claims that they evolve cripplingly expensive ornaments to demonstrate to potential spouses that their genes are good enough to bear the cost. This is, at least, an amusing idea. It has been used to explain bizarre patterns of human behavior such as drug abuse. Perhaps men take alcohol, tobacco or stronger drugs to demonstrate to women how tough they are, how their constitutions can cope with mistreatment and how they might make excellent fathers as a result. One of the most baffling findings of modern anthropology is the discovery of small tubes in the tombs of Maya Indians. Handicappers believe that these were used to give ritual enemas of toxic drugs to the most powerful men, guaranteeing instant intoxication and a widely admired statement of sexual prowess. The habit has not yet spread to the streets of New York.

Conflict between males for the attention of females is often obvious and may be painfully close to the experience of the biologists who study it. There are also plenty of chances for conflict between males and females. In some animals, the reluctance of females to accept a new mate, however persistent he might be, arises because males invest less in bringing up offspring. It pays them to mate and run; to try and father as many children with as many females as possible. Females need to be more cautious. As it costs so much to produce and bring up a child they may choose the male who is likely to be the best father and reject the rest.

This conflict of interest is sometimes brutally obvious. In many creatures males kill a mother's offspring by another male with the aim of making her available to themselves. In the Langur Monkeys of India, most of the young die for this reason. There is even a form of prenatal cannibalism. Pregnant female mice and horses exposed to a new male reabsorb their fetuses, a behavior which may have evolved because of the near certainty that if the offspring are born they will be killed.

Humans reveal the intersexual conflict in less blatant ways. Their battle is an economic rather than a mortal one. If tribal peoples are any guide, there is more polygamy in societies which have invented private property, as women prefer well-endowed mates. When wealth is concentrated into few hands life becomes more like that of a gorilla, with the richest males monopolizing the females. For example, the philoprogenitive (and opulent) Moulay Ismail the Bloodthirsty of Morocco admitted to 888 children. Although the West now seems to be moving toward the chimpanzees, with most men having at least a chance of finding Ms. Right, in some societies mating success is still related to wealth. Among the Kipsigis people of southwest Kenya women prefer rich husbands. A wealthy man may have as many as a dozen wives and eighty children and the more land a man has the more wives he is likely to possess. Many of the poorest males leave the community as teenagers and have no children at all. The women nearly all have families of about the same size. There is an economic conflict between the sexes with men providing the capital and women choosing where to invest. In Western countries, too, men from higher social groups have many more partners than do those who are less well off.

When one people conquers another it is the men who capitalize on their dominant position to find new mates. In the "Cape Colored" population of South Africa (which is intermediate in appearance between Africans and Europeans) most of the genes are halfway between those of modern black and white South Africans. However, nearly all the DNA of their Y chromosomes are of the European type, showing how white males took advantage of their economic domination over black females in earlier centuries. Exactly the same happened in the Americas. Many of the Y chromosomes of modern native Americans in Mexico populations are of European origin.

A battle between the sexes may help to explain another unusual attribute of human reproduction. Women are the only female primates who do not make it obvious when they are most fertile. In dogs and many other mammals the female goes "into heat." Most female primates advertise the two or three days in each cycle when they are most likely to conceive. Often, this is accompanied by a frenzy of copulation with a series of males. Before modern medicine, most women (and all men) were unaware of when the fertile period was. Women's reproductive coyness may perhaps reflect the change in the economic relation of the sexes which came with the beginnings of society. It might be an attempt to resolve the conflict between male promiscuity and the female's need to ensure the care of her offspring. By concealing when she is fertile she ensures constant attention from her mate. If he is not sure when she can conceive then he dare not leave her for a new woman in case another male takes advantage of his absence. This is, of course, historical speculation with no direct evidence for or against it.

Males do, needless to say, contribute to the care of their children. However, there is in most societies a difference between the sexes in their commitment. If a relationship breaks up it is usually the mother who is left holding the baby. The difference can be seen in subtle ways. There are many genetic tests which tell parents whether they carry a harmful gene and hence whether it is wise for them to plan to have children. In a few cases, the test also tells the parents themselves that they are at risk of developing the disease later in life. Huntington's Disease is of this kind. Twice as many women as men volunteer for testing, perhaps because their concern for their potential child's future is greater than that for their own peace of mind.

The battle of the sexes is usually seen as regrettable but unavoidable. There is a natural tendency to assume that the bonds between mother and child are driven by mutual devotion. However, to the cold eye of the biologist the transaction between generations is also based on conflict. There are many chances for mother and child to exploit each other. It is in the child's interest to gain as much attention as possible from its mother. The mother's concern is to provide as little as will allow her offspring to survive. If she is too generous to one child, the next may suffer.

Such confrontations, shocking though they may seem, are the commonplace of the animal world. There has grown up in biology the comforting supposition that nature is not really red in tooth and claw and that animals rarely do much harm to other members of their own species. The battle for reproductive success shows how wrong this is. Eagles lay several eggs. If food is plentiful all the chicks are fed; but if there is a shortage then the last to hatch is allowed to starve or is killed by its siblings. Rats, mice and other mammals often eat all their young when food gets short and there is even a word for this—kronism, after the Greek deity Kronos, who devoured his own children.

Any mother is certain, of course, that all her children (first, second or third born) carry her own genes. However, it is quite possible (and in many animals almost guaranteed) that the father of her first child will not be the same as that of her later offspring. As Aristotle put it in the fourth century B.C., "This is the reason why mothers are more devoted to their children than fathers: it is that they suffer more in giving them birth and are more certain that they are their own." The conflicts of interest involved, and the differences in the amount of investment of each sex in their offspring, lead to some subtleties in the battle of the sexes and of the generations and may help to explain some rather strange patterns of inheritance.

Much to the surprise of geneticists the effects of a particular gene sometimes seem to depend on whether it is passed on by mother or by father. This effect, "genomic imprinting" as it is known, is quite different from sex-linkage (see p. 51), and the genes involved may be on any chromosome. Each sex seems to stamp its personality on the copy of the gene which it transmits. Although the DNA itself is not permanently altered, its effects on those who inherit it depend on which parent it came from. A gene passed on by a father to his daughter differs in its impact from those of the same one when she passes it on to her own children. The DNA is "marked" as it is conveyed through sperm or egg and the mark is reversed whenever the line of transmission changes from one sex to another.

The effects of imprinting can be seen in the inheritance of Huntington's Disease. The age at which the symptoms of nerve damage first appear varies from person to person. Those who inherit the gene from their father show its effects sooner than do those receiving a copy of the same thing from their mother. The children of affected men show the first symptoms at an average age of thirty-three while children whose mother has the disease are healthy for another nine years. The influence of the gene (but not the gene itself) is modified as it passes through sperm or egg.

Each developing embryo contains, of course, both maternal and paternal DNA. If we use the (rather dubious) metaphor that each gene acts in its own interests, it pays those coming from the father to extract as much as possible from the mother in which they find themselves, irrespective of any damage which this does to her and hence to any subsequent children. This is because later offspring will probably carry a separate set of genes from a different father. The first father loses nothing by exploiting his mate as much as he possibly can. The mother, in contrast, needs to ensure that further attempts to pass on her own biological heritage are not jeopardized by the avarice of her firstborn. This may explain the difference in behavior of the same gene when it is transmitted through fathers or through mothers.

There is some evidence for the idea that imprinting arises from paternal greed and irresponsibility. In mice, the genes which produce the membranes through which the developing fetus feeds are more active if they come from the father than from the mother. Those passing through the father also tend to increase the size of the tongue which is, of course, used in suckling. Genes for human disease show the same effect. Some fetuses accidentally inherit two copies of a gene promoting growth. They grow abnormally large only if both copies come from the father. In normal fetuses only the paternal copy is switched on, again showing the father's interest in his child extracting the maximum nutrition from its mother. Two rare genetic diseases (which glory in the name of the Prader-Willi and the Angelman syndromes) were once thought to be different as their symptoms are so distinct. In fact they are due to the same mutation. The differences depend on whether it is passed on by father or by mother. Children with Prader-Willi syndrome (whose abnormal gene comes from their father) suckle greedily and are fat, while Angelman children (who receive the same gene from their mother) are of normal weight.

Enthusiasts for conflict suggest that even a baby's crying at night is an attempt to manipulate its mother to provide more food and that the mother retaliates by secreting in her milk substances similar to those used as sedatives by doctors. Whether or not this is true, it is clear that once sex has evolved it has some unexpected effects on the lives of the creatures who practice it. Without sex there would be almost no evolution and no genetics. Our universal fascination with the subject may, one day, provide the answer to the most important sexual problem of all—why we bother in the first place.

6 | Clocks, Fossils and Apes

THE BOUNDARY between apes and humans was once far from clear. Lord Monboddo, a friend of Dr. Johnson's, was convinced that "The Orang Utan is as ardent for women as it is for its own females" and that the Malayans cut the tails off the offspring of such matings and took them as their own. "From the particulars mentioned"—he wrote—"it appears certain that they are of our species . . . though they have not come to the lengths of language." Dr. Johnson was not impressed: "It is a pity to see Lord Monboddo publish such notions . . . in a fool doing it, we should only laugh; but when a wise man does it, we are sorry."

There is a complementary foolishness around today. In the United States, four people in ten do not believe that humans are related to apes at all or even that the human species is more than a few thousand years old. Creationists are determined to stay ignorant. They deny that we evolved and are hence connected through our genes to the rest of the living world. In 1982, President Reagan said that "Evolution is only a theory which is not believed in the scientific community to be as infallible as it once was . . . Recent discoveries have pointed up great flaws in it." The creationist dogma bores, when it does not exasperate, biologists. As a result they have been less active in fighting it than they should and the bigots have had some success, in the U.S.A. at least, in forcing their views onto children.

The best of all evidence that humans did evolve and that they are members of the animal world as a whole comes from fossils. To most people, the study of evolution is the study of fossils. Without them, the portrait of our forebears can never be complete. Place yourself in the position of a historian who knows only about the modern world. To infer the progress of, say, Turkey and the United States just from what exists today would be almost impossible. Historians need documents from the past. To have any real confidence in their theories evolutionists must have the same thing.

The written documents of history stop—effectively—at the day before yesterday. The earliest texts come from the Sumerians. Records go back a little further in mythic form. Gilgamesh was King of the Sumerian city-state of Uruk in 2700 B.C. The epic which bears his name has some familiar features. There is a Garden of Eden, a hero's descent to the Underworld (with a safe return) and a flood. Excavations in the Middle East show that there was indeed a gigantic flood at about that time. Even legend retains some fragments of the truth.

Fossils authenticate the past. At one time, they were regarded as such powerful evidence by creationists that they were defined to be the work of the Devil, placed in the rock to mislead the faithful into believing in evolution. Later, there was a last-ditch attempt to fit them into the Bible. Some fossil mammals seemed to be standing on tiptoe with their noses in the air when they met their end. Obviously, they had been overwhelmed by Noah's Flood.

Darwin was well aware of the power of the relics of the past in supporting the idea of evolution. About one page in six of his first book, *The Origin of Species,* deals with the fossil record of animals and plants. The fragments of their ancestors were central to his theory. Even for animals, he recognized that the record was very incomplete: ". . . a history of the world, imperfectly kept, and written in a changing dialect; of this history we possess the last volume alone . . . of this volume only here and there a short chapter . . . and of each page only here and there a few lines."

For humans, Darwin had an enormous gap in his evidence. He knew nothing about the remains of our ancestors and there is scarcely a mention of them in his other great work, *The Descent of Man,* published in 1871. Although we now know a little more about the bones of our predecessors the record of our own evolution is still very incomplete.

The first fossil to be recognized as a human ancestor was Neanderthal man, found in the Neander Valley in Germany in 1856. Such was the power of belief in those days that some dismissed the bones as those of an arthritic cripple or of a cossack who had died during the

retreat from Moscow. A hundred years ago, a skull intermediate between humans and other primates was found. This was *Pithecanthropus
erectus*, Java Man. The search for our birthplace and our migration
routes was on and has continued ever since.

Paleontologists still do not agree about where modern humans came
from and where they went. The fossil record is so incomplete that a
cynic might feel that the main lesson to be learned from it is that
evolution usually takes place somewhere else. The origin of humanity
has been claimed as being in Asia, Africa and even the whole world at
the same time. The human record has been investigated as intensively
as any, but there are still enormous holes in it. Even the best known
deposits are very incomplete. The area around Lake Turkana in East
Africa is almost never off the television screen. Guesses about population size from the food available suggest that perhaps seventy million
people lived there over its two-and-a-half-million-year history. Remains of only about two hundred have been found, mostly as small
fragments. The fossil record will never provide the complete history of
human evolution, but it can give dates and places which genes can only
hint at. It is worth glancing at the bones before staring at the molecules.

Just as is the case with the genome, the biggest problem in getting to
grips with the preserved record of the past is that of scale. Life began
about three thousand million years ago. We can use the journey from
Palm Beach to the southern tip of Manhattan up the Eastern Seaboard
as a metaphor for its history (it has, after all, already served to illustrate
the size of the human genome). Florida, Georgia and most of South
Carolina are covered by primeval slime about which we know nothing.
The first primitive land animals crawl ashore near Norfolk, Virginia.
There are frogs in Delaware and the streets of Atlantic City are in-
fested with dinosaurs. Early primates appear on the north coast of New
Jersey, and our own species can look over the polluted waters of the
Hudson River from its birthplace a hundred yards from the Battery.
Recorded history begins on the shore, at high tide mark.

The journey needs milestones. As it is a voyage through time, they
must mark the important dates in history. There are many ways of
dating fossils. Some depend on the decay of radioactive materials as
time passes. Others are more ingenious. Ostrich eggs were favored as
containers in the ancient world. The structure of their amino acids, like
that in all living tissues, is biased toward the left. Over the years, the
amino acids decay into a mixture of left- and right-handed forms. Mea-

suring the ratio of left to right dates the shells and the people who used them. The oldest known ostrich-shell containers are at Klasies River Mouth, a site in South Africa occupied by humans whose skulls look much like those of today. They are dated by the ostriches as a hundred and twenty thousand years old. The oldest outside Africa, found in the Israeli cave of Qafzeh, are twenty thousand years younger; and fifty thousand years ago the shells were used to make the first of all ornaments, some beads unearthed in Tanzania.

The history of mankind's earliest ancestors is obscure. Bones that look like those of primates—apes, monkeys and humans—appear around sixty million years ago. The first fragment of an anthropoid (the group which evolved into monkeys, apes and humans) is about fifty million years old, from Algeria. This creature was not much bigger than a rat. One jawbone half that age from an early hominoid (the group which includes humans and apes) has been found in Kenya. It is the sole relic of history in a ten-million-year period which covers the split of the line leading to monkeys from that to apes and ourselves. By fifteen million years ago several species of ape were roaming Africa and Asia. None was larger than a seven-year-old boy and all had small brains and pointed faces. There is then another ten-million-year gap in the record. A fossil face from this period turned up in Macedonia in 1991. This ancient Greek is the nearest thing yet found to a common ancestor of apes and humans.

What seems to be the earliest direct precursor of modern humans appeared between three and four million years before today in the Laetoli beds of Kenya. This creature, *Australopithecus afarensis*, is named after the Afar region of Ethiopia, the Biblical Ophir referred to in the story of Solomon and the Queen of Sheba. The most famous specimen is "Lucy," so named because the discoverers were playing the Beatles' "Lucy in the Sky with Diamonds" at the time. She was less than four feet tall, with a small skull and a slouching gait. The earliest bones which look as if they belong to our immediate ancestors, the genus *Homo*, come from Kenya and are dated at about two and a half million years old. The first stone tools appear at about the same time.

It is difficult to classify fossils in the same way as living organisms. The problem is rather like defining artistic styles. Because they evolve one into another it is meaningless to draw a line showing exactly when, for example, impressionist painting changed into post-impressionism. A certain arbitrariness is bound to creep in. In paleontology things are even worse, as only a few specimens are found and there is a natural tendency to grace each with its own name. Most paleontologists agree that there were three (or possibly four) species of *Homo:* the first, *Homo habilis,* ("handy man") from over two million years ago; the second, *Homo erectus,* rather more recently; and—finally—our own species, *Homo sapiens,* which began to emerge about half a million years ago. *Homo habilis* is sometimes divided into two distinct species, *habilis* itself and *Homo rudolfensis. Habilis* had a larger brain than its predecessors, its face jutted out less and for the first time there was a noticeable nose and a perceptible chin. An almost complete skeleton of a *Homo erectus* boy has been found near Lake Turkana in Kenya. He had a protruding brow ridge and a massive jaw with long arms and legs. For much of the time more than one species of man-like creature existed at once. In Africa beasts which looked much like Lucy and her relatives lived for thousands of years alongside *Homo habilis.* Later, there may have been two coexisting species of *Homo,* a situation which would be intriguing if it held today. In fact, humans are unusual among mammals in being the sole extant member of their evolutionary family, the genus *Homo.* They share this distinction only with the aardvark.

Homo erectus was the first to escape from Africa and did so soon after evolving. An *erectus* jawbone, mixed with the bones of saber-toothed tigers and elephants, is preserved beneath the town of Dmanisi in Caucasian Georgia. It may be as much as 1.8 million years old. Within a million or so years *Homo erectus* had spread to the Middle East, China, Java and Europe. "Java Man" and "Peking Man" (whose bones mysteriously disappeared during the chaos of the Japanese invasion of China) both belonged to this species, which had a static way of life with almost no change in the skull over its long history.

Early *Homo sapiens*—some of which look rather like *erectus*—emerged in Africa around four hundred thousand years ago. These creatures would appear distinctly threatening to modern eyes, although some had brains larger than the average today. Within a couple of hundred thousand years there was a European population of "archaic *Homo sapiens*" (to which the first Briton, Swanscombe Man, belonged). They may have evolved into the Neanderthals, whose remains have turned up all over Europe and the Middle East.

The Neanderthals themselves, who flourished for a hundred thousand years before disappearing before a wave of modern humans, had larger brains than our own (albeit on a heavier body frame), with large noses and teeth. Their arms and legs were short, rather like today's

Eskimos and like them they were adapted to living in the cold. Neanderthals have been found as far east as Iraq, but not in Africa or elsewhere.

Around a hundred and thirty thousand years ago, the first humans of distinctly modern appearance (light build, thin skull, large brain and small jaw) appear in Africa. Their remains have been found in Omo-Kibish in Ethiopia. Next to the overhanging cliff at the South African site of Klasies River Mouth, in fifty feet of sediment which encompasses forty thousand years of history, there are many fossils of this type. Empty shells are scattered about. The food—which seems from its remains to have come mainly from the sea—was cooked on open fires. These early modern humans had arrived in Israel, in the caves of Qafzeh and Skhul, by a hundred thousand years ago. Cro-Magnon man, the first modern European (who lived, like a sensible man, in the south of France) was there by forty thousand years before the present day.

This account of history is the "out of Africa" model believed by most evolutionists. However, there is another theory. Some feel that humans emerged more or less at the same time over the whole world so that today's Chinese evolved from an ancient Chinese ancestor and Africans from a predecessor in their own land. The idea that the same species can evolve simultaneously in different places flies in the face of theories of the genetics of speciation (which is not to say that it is wrong). Some fossils might support the idea of local evolution. One, found in 1990 near the Han River in China, resembles *Homo erectus*, but has a flattened face which, to its discoverers, looks rather like that of a modern Chinese. Those in favor of local evolution make much of the "shovel incisors" in fossil jaws from Asia. The teeth are scooped out at the back, as are those of some of today's Chinese. In some places in Europe up to a third of people have shovel incisors, so that this is not a very convincing argument. So few early fragments have been preserved that it seems that too often history is in the eye of the beholder. Africa was the world center in which most primates originated and there is no reason to suppose that humans were any different.

Another fossil controversy which has received more publicity than it deserves is the question of evolution by creeps or by jerks. Darwin felt strongly that the origin of species was a gradual and continuous process. The past was no more than the present writ large. Because of the vast length of time available the enormous transformations which took place through the history of life can, he thought, be explained by the slow and almost imperceptible changes which influence living creatures today. His was a leisurely and Victorian view of the way the world worked; one of gradual and almost inevitable progress.

There is an opposing view (the theory of "punctuated equilibrium" as it is known in its latest guise) which has a more twentieth-century flavor. It sees evolution as boredom mitigated by panic. New species appear during a sudden burst of revolutionary change. Between these historical disasters, life is tranquil. In spite of Darwin, the punctuationists claim, the origin of species has nothing to do with what happens to a species once it has originated. Looking at the process of evolution today cannot tell us much about what went on in the past.

The greatest strength of this theory is its ability to annoy Darwinists. Hundreds of scientific papers have been written in support of or against punctuated equilibria. One important problem is that of timescale. What might appear an instant to a geologist can seem an eternity to a biologist. A "jerk" between one species and its successor may encompass tens of thousands of years: nothing in terms of geological epochs, but more than enough generations to allow plenty of gradual evolution of the Darwinian kind to make fundamental changes. Those opposed to creeping evolution point out, quite fairly, that most species do not change at all through their evolutionary lifetimes, which is not what Darwin would have expected.

Whatever the merits of each doctrine there are so many gaps in the human fossil record that there is just not enough information to tell whether humans evolved suddenly or slowly. The remains are so sparse that it is quite possible that relics of the lineage leading to the peoples of today are as yet undiscovered. The unfortunate truth is that although fossils are the best of all evidence that we did evolve, they cannot tell us much about how that evolution took place. What is clear is that the attributes which make us what we are arose bit by bit, appearing first in a remote ancestor and reaching completion (if indeed they have) only in the past hundred thousand years or so. No single primate awoke one morning to find itself human.

The most important problem in using preserved remains to study history is that it is impossible to be sure that any fossil left a descendant. Reconstructing human evolution from their fragments can never be much more than clasping at straws. Our extinct predecessors are just that: extinct. This makes it difficult to work out their relationships to each other and to ourselves.

However, there is another window onto the past. Every modern gene descends from times long gone. The connections between humans and primates are preserved in the DNA of living animals. Darwin himself realized that there are better ways of looking at history than depending on the frozen accidents which are fossils. All his claims about the predecessors of humankind depended on indirect evidence (such as comparing the anatomy of humans with that of apes). Nowadays this evidence is much more complete and a portrait of our ancestors is beginning to emerge.

Molecular biology is just anatomy writ small, plus an enormous research grant. To a geneticist, everyone is a living fossil, containing the heritage of his or her predecessors. Genes re-create history, not just since humans appeared on earth, but since the origin of life itself. *The Descent of Man* compares humans with monkeys and apes to establish their common ancestry. "Man"—Darwin said—"still bears in his bodily frame the indelible stamp of his lowly origin." W. S. Gilbert put it more wittily: "Darwininian man, though well-behaved, is really just a monkey shaved." Genetics allow us to search for who that shaved monkey may have been, and even when it lived, by looking at our relatives.

Bones show that humans are more closely related to apes than to monkeys and that their closest kin lie among the chimps, gorillas and orangutan. Anatomists once assumed that *Homo sapiens* must be quite distinct. Often, it was contrasted with these "great apes." Humans differ from them in obvious ways—brain size and hairiness, for example—and have some other unique talents. Most people are right-handed and, judging from the patterns of breakage of stone tools, so were our ancestors. Although individual chimps and gorillas may use one hand rather than the other, about half the animals prefer left and half right. The human brain, too, is asymmetrical and it may be more than a coincidence that speech and language are coded for on only one side.

As it is hard to measure how much genetic divergence a difference in hairiness or handedness represents, such comparisons are not very useful in measuring the biological gap between apes and humans. Genetics can do a better job. We share many genes with apes. Not only do we both vary in the way we taste the world (see p. 40), but we share diversity in how we see it. In some monkeys, many of the males are red-green color-blind. Chimps have A and O blood groups, while gorillas are all blood group B. About a thousand distinct stained bands can be seen in the human chromosome set. Every one is also found in chimps. The main change is not in the amount of chromosomal material but in its order. Many of the bands have been reshuffled and two chromosomes are fused together in the line leading to humans. We have forty-six chromosomes in each cell, while chimps and gorillas have forty-eight.

At the DNA level, too, there has been remarkably little change. In one of the hemoglobin pseudogenes, the "rusting hulk" of a working gene (see p. 65), which accumulates mutations quickly as it has no function, humans are about 1.7 percent distant from both chimps and gorillas, 3.5 percent from orangutan and 7.9 percent from rhesus monkeys. Other genes suggest a closer link between chimp and gorilla, hinting that the line leading to humans diverged earlier than the chimp-gorilla branch. Chimps as a species are more variable at the molecular level than are humans, suggesting that they have been evolving in the same place for longer.

To sort out man's place in nature we need to look at as many genes as possible and to combine information from them all. A new method known as DNA hybridization does exactly this. It depends on the extraordinary toughness of the DNA molecule and its overwhelming desire for togetherness; that is, for each strand to pair with a sequence which matches its own.

When a DNA double helix is heated up it separates into two single strands, each bearing a matching set of the four bases. As the liquid cools, the strands come together, A pairing with T and G with C, to reconstitute the original double structure. If DNA from two different species is treated in this way the same thing happens. Some of the single strands from each one form a hybrid molecule containing one strand from either species. The more closely related the two species the more similar their DNA and the tighter the fit. If the strands are very similar, they stay together even at high temperatures, but if they share fewer DNA sequences they are less stable. The temperature at which the hybrid melts hence estimates how alike the two DNA sequences are. This is a quick and easy way of measuring the relatedness of any pair of species. It has already sorted out some thorny problems of classification. For example, DNA hybridization shows that the nearest relatives of New World vultures (which include the California Condor) are storks, rather than the vultures of the Old World.

The results from primates are surprising. Humans and chimps share 98.4 percent of their DNA, slightly more than either does with the gorilla. The orangutan is less closely related and New World monkeys even less so. Any idea that humans are on a lofty genetic pinnacle is simply wrong. A taxonomist from Mars armed with a DNA hybridization machine would classify humans, gorillas and chimpanzees as members of the same closely related biological family.

This certainly does not mean that humans and chimps are just minor variants on the same theme. There is more to evolution than change in DNA. The Hawaiian islands have more species of fruit fly than anywhere else in the world, with a vast diversity of form. One looks like a hammerhead shark, with huge protuberances on either side of its head. DNA hybridization shows that this wild evolutionary euphoria is accompanied by almost no change in the genetic material. The same is true of another bizarre group of creatures, the cichlid fishes of East African lakes. There are hundreds of species, doing things as different as grazing on algae, preying on each other and even adopting the rather disgusting life-style of browsing on the scales of their living fellows. If fish were mammals we would perceive them to be as different as deer, wolves and rats. At the DNA level they are almost indistinguishable. The group of snails upon which I myself work, on the other hand, look much the same, but each species has very different genes. For some reason, evolution of their body form has been sluggish while their molecules change quickly.

Our brains and our behavior are what separate humans from any other animal. They probably involve a few genes whose importance is lost in a measure of average genetic difference. There is also, of course, a whole set of intellectual and cultural attributes which appear once a crucial level of intelligence has been reached and which are not coded for by genes at all. The human brain is three times larger than would be expected for a typical primate of the same size—and our two-million-year-old ancestors were already in the forefront of their primate relatives in this regard.

Somewhere in that brain, or what it is thinking, is what makes us different. Although it shares most of its DNA with humans, no chimp can speak. There are claims that they can manipulate symbols in a primitive sort of "language" (although trained parrots can do almost as well). The attempt to show that apes might talk is one of the great blind alleys of behavioral research. Samuel Butler's comment on a Victorian attempt to teach a dog sign language is worth remembering: "If I was his dog, and he taught me, the first thing I should tell him is that he is a damned fool!" To make too much of the shared DNA of chimps and humans is to be in danger of falling into the same foolishness. Humans, uniquely, are what they think.

Whatever its limitations, gene sharing can say a lot about history. All the biological differences between humans and their relatives come from mutations, genetic accidents since the primates began to diverge. They can be used to guess at when the human family tree separated from the others: the more differences, the longer ago the split. If mutations happen at a regular rate, they can even be used as a "molecular clock," which uses changes in genes to infer when two stocks last shared a common ancestor.

Molecular clocks depend on a number of assumptions, some of which may even be justified. First, mutations must happen at a constant pace as one generation succeeds another. In addition, they should have few effects on those who carry them and, as most happen in parts of the DNA which do not contain any meaningful instructions, perhaps they do not. DNA errors accumulate over the years. Although some are lost because, by chance, those who carry them do not reproduce, these are replaced as new mutations come along. The genetic makeup of any lineage hence changes with time. The transformation of the inherited message in related species gives a clue as to when they began to diverge. To date the split, there must be evidence from fossils (or from other sources such as the date of appearance of a barrier like a mountain range) as to roughly when two living members of the group being studied last shared a common ancestor. Comparing their genes sets the rate at which the clock ticks and makes it possible to work out the date of separation of other species whose ancestors left no fossils of their own.

Linguists use the same logic as biologists to unravel the history of the world's tongues. As words are passed from parents to children errors creep in. Sometimes the changes are scarcely noticeable. In Shakespeare's *As You Like It* the court jester makes a speech which causes great amusement: looking at a clock he says, "Thus we may see how the world wags; 'tis but an hour ago since it was nine; and after one hour more 'twill be eleven; and so, from hour to hour, we ripe and ripe; and then, from hour to hour, we rot and rot; and thereby hangs a tale." Just why this should be so funny is lost on modern audiences—unless they realize that in Shakespeare's time the word "hour" sounded almost the same as the word "whore." Such errors can be used to date manuscripts. Before the invention of printing, texts were copied by hand, often by people who had little idea what they meant. As copy followed copy, more and more mistakes crept in. Counting the inaccuracies gives a very good idea as to just when a particular version of an ancient original was in fact written.

Such changes are small, but can make a big difference. Languages as distinct as Bengali and English are related. They owe their existence to the accumulation of tiny modifications of a common ancestor spoken long ago. Take the word for kings and queens. In Sanskrit this was *raj*, in Latin *rex*, in Old Irish *ri*, in French *roi*, in Spanish *rey* and in English *royal*. There have been different transmission errors in the pathway leading to each language. If we know the date of the split (as we do from literary fossils) we can make a linguistic clock. In Europe, this ticks at a rate which means that two languages share about eighty percent of their words a thousand years after they divide. The language clock is an imperfect one: some words scarcely change while others shift more quickly. Nevertheless, it can be used to trace the origin of modern tongues although their ancestral speakers are long dead.

The idea of a molecular clock driven by mutation is beautifully simple. As usual, the more we learn the worse it gets. It speeds up and slows down, and ticks at different rates for different genes. The same confusion led the nineteenth-century Linguistic Society of Paris to ban discussion of the origin of languages. There have been some spectacular failures by molecular clockmakers to see the biological wood for the evolutionary trees. However, they have had some triumphs. One of these is, by sad accident, bound up with the story of hemophilia.

In the most common form of hemophilia the factor VIII gene is not working properly. A few years ago, it became possible to treat hemophiliacs with factor VIII from donated blood. In Britain, this was bought from the United States. The American blood donation system is a commercial one. Many of those involved were drug addicts who sold their blood, some of which was contaminated with the AIDS virus. In the U.S., about fifteen thousand hemophiliacs treated in this way were infected. The number is smaller in Britain, but some have already died. The clock can give an answer to the universal question—where did AIDS come from and when it did start to infect us? There have been a number of wild theories—for example that a monkey virus infected human polio vaccines thirty years ago and led to the AIDS epidemic. The truth is less sensational but may be equally alarming.

The AIDS virus is tiny, with a genetic message of only about ten thousand letters. The first American cases began to appear in the late 1970s. The disease was probably around before then as the genes in a pickled specimen from a sailor who died in Manchester in 1959 show that his death, then a mystery, was in fact due to AIDS. Since then, millions of people have been infected.

The virus reproduces so quickly that its evolution can be studied over quite short periods. Its DNA sequence can even change within a single person during the course of infection. The speed of change— and the way in which genetics can be used to reconstruct evolution—is illustrated in the tale of a Florida dentist, David Acer, who died of AIDS. One of his patients, Kimberley Bergalis, who had no apparent reason to be at risk, herself died from the disease. Then, four more of his clients were diagnosed as having it. There were claims that the patients had caught the disease from their dentist's blood, perhaps because he had a small cut. One, Richard Driskill, sued the dentist's insurance company for fifteen million dollars. The company argued in court that the infection was his fault as he had been promiscuous and used drugs. DNA and the molecular clock gave the answer.

The DNA of the AIDS viruses isolated from Acer and his patients showed that they were similar (but not identical) to each other and quite different from samples of the virus taken from other Florida patients. In spite of some statistical doubts, the insurance company settled out of court for an undisclosed sum. The evidence from the molecular clock that the dentist's virus was the immediate ancestor of the one that infected Driskill was strong enough to convince them that there was no point in fighting the case. Genetics had, at least to their satisfaction, re-created history.

Different parts of the virus' DNA evolve at different speeds. By choosing a section whose clock ticks at the appropriate rate, it is possible to estimate when virus strains which are now very different split from each other. If DNA from patients in different parts of the world is used to make a tree of relatedness, it seems that the ancestral virus came from Africa and spread via Haiti to the United States and then to Europe. Because the date of the first infection in each place is known we can set the rate at which this part of the AIDS molecular clock ticks. As many as fifty mutations accumulate in each lineage each year.

Comparisons of AIDS virus genes with those of others shows that it is related to some which infect primates. They are widespread in African monkeys, but not in those of Asia or the New World. Its closest relative attacks the sooty mangabey, a West African monkey. The molecular clock based on comparing genes in AIDS viruses with those in the monkey virus suggests that the most important agent of infection split off from its ancestor and invaded humans well over a century ago —probably on more than one occasion. Why, nobody knows.

The AIDS clock helped to unravel the history of an organism which evolved in the recent past. We are a lot less confident when using the same approach to date human evolution. The main problem is that the fossil record is so patchy that it is difficult to find firm dates—like, say, the first appearance of AIDS in the U.S.A.—with which to set the clock. Fossils suggest that the line leading to baboons had split off by twenty-five to thirty million years ago and that to orangutans by twelve to sixteen million years before the present. They say nothing about the date of the split between humans, chimps and gorillas. A molecular clock based on the genes of our primate relatives suggests that this split happened six to eight million years ago, with the gorilla line breaking off just before the divergence of those leading to chimps and to humans. For some reason, the clock has slowed down in the line leading to humans compared to that terminating in the New World Monkeys.

The last common ancestor of chimps and humans lived about three hundred and fifty thousand human generations ago. This is less than the number of generations separating the AIDS virus of today from that of its primate predecessor. *Homo sapiens* is a recent arrival even in the history of the primates, let alone in the three-thousand-million-year pedigree of life itself.

When—or how—the attributes which separate humans so absolutely from any other creature appeared is not a question for biology. Perhaps the best we can do is to agree with Keats that we are all "twixt ape and Plato" and leave it to individual preference just where on that long road we place ourselves.

7 | Time and Chance

THE GOOD BOOK points out, in Ecclesiastes, that "The race is not to the swift, nor the battle to the strong . . . but time and chance happeneth to them all." Evolution is, of course, all about change and time, but how things evolve is often a matter of chance. Much of the human condition is shaped by accident. The nature of inheritance means that random events are bound to direct our genes as the generations succeed one another.

The importance of accident in evolution was noticed by the English cleric Thomas Malthus (who is better known for giving Darwin the idea of the struggle for existence). He became interested in the history of the burghers of Berne and followed their family names over several centuries. To his surprise, many of the surnames present at the beginning of the period had gone by the end, although the number of burghers stayed about the same. Francis Galton showed why.

A surname is rather like a gene; it passes from father to son. Every generation there is a chance that a particular father does not have a son. Perhaps he has just daughters (who lose their names on marriage) or no children at all. His name is then lost from the family line. If he has no children the name will certainly disappear. It is also more likely to become extinct if he has a small family, as a brood of one or two children will quite often consist only of daughters. If, in a closed community like that of the Bernese bourgeoisie, this process continues for long enough, more and more names will disappear as the years pass. In theory, and given enough time, only one surname will survive, although the number of people involved need not change. The population will then be homogeneous. Everyone will carry the same inherited message (at least as far as their name is concerned). The community will also be more inbred than before, as the only mates available will be people sharing the same family name, all of whom descend from a common ancestor.

Exactly the same thing happens to genes. Perhaps, among the Bernese bourgeoisie, there was a rare gene—possibly an unusual blood group. Because Berne was a small town, only a few people carried it. If none of them passed it on (because they had no children, or because by chance the gene did not get into sperm or egg) then it was lost. On the other hand, the carriers might, again by chance, have had more children than the rest, in which case the variant became more common. In either case, its frequency altered (which means that the population evolved) but the change was purely accidental.

Einstein once said that "God does not play dice." He was wrong: for genes, God does. As when throwing dice, what number comes up has nothing to do with the properties of the DNA involved. In some ways this is a deeply theological issue. Is it their own fault that genes, and those who carry them, are damned—or do they perish at random because of simple bad luck?

Just as for surnames, random genetic change is more likely to take place in small populations, when only a few people bear a particular gene. In these circumstances, all or most of the carriers may, by chance, fail to pass it on. In a larger group, although a variant may still be rare it will be borne by enough people to make it likely that at least one will transmit it.

Such evolution by accident is known as genetic drift. It has certainly been important in our own past. *Homo sapiens* was until recently a rare species, living in small bands. Until a few tens of thousands of years ago there were no more people worldwide than live in London or New York today. There are hints of what human society was like for most of the past in the few tribal peoples who have survived.

Until a few years ago, when their lives were destroyed by mining and logging, about ten thousand Yanomamo Indians lived in a hundred scattered villages in the rainforests of southern Venezuela and northern Brazil. They referred to themselves as "the fierce people," with good reason. About a third of all male deaths were due to violence, often during battles between the villages and, the Yanomamo believed, many more to malevolent magic worked by other villages.

Their society was not robust enough to allow groups of more than eighty to a hundred people, including around a dozen young adult males, to stay together. Any larger band tended to split. The splinter group moved away to found a village somewhere else. The Yanomamo existed for their whole history (which stretched back in some form to the peopling of the Americas twelve thousand years ago and more) as a series of small communities in constant conflict.

Social systems based on hunting and gathering, as all were for nine tenths of our evolution, may each have been rather like this. The ancient Siberians who hunted mammoths made houses from their bones. Measuring the size of their bony villages suggests that each group, like today's Yanomamo, consisted of a few score people. It is dangerous to make too much of what one tribal culture like the Yanomamo does. Others, such as the Bushmen, are much better behaved. But there is one odd fact about modern society which may be trying to say something about the size of ancient social groups. This is that most team efforts involve about the same number of individuals. There are nine members of the U.S. Supreme Court, eleven on a football team, twelve on a jury—and Jesus, of course, had twelve disciples. Each Yanomamo band has, curiously enough, about a dozen healthy adult males. Is the difficulty in reaching consensus in a larger group a hint about society during much of history? In the United States most people can identify about twelve others whose death would cause them anguish. Aristotle himself pointed out that it is impossible to love more than a few. Could all this be a clue (albeit a feeble one) about the size of communities in the distant past?

Strange things befall genes in small populations such as these. Again, surnames show what can happen. The evolution is easy to study (needing only a telephone book) and names are preserved for centuries in marriage records. There are about a million surnames in the world. Those in China are the oldest, dating back to the Han dynasty two thousand years ago. In contrast, Japanese surnames go back only a century or so, when names were ascribed by order of the authorities. There are various complications in using them. For example, in many places the same name (like my own, Jones) appeared independently many times; in my case it merely means "son of John." In some societies, such as those of Spain and Russia, the whole system breaks down as children take the name of their father and the "surname" changes each generation. The same was once true in Wales. A boy would take the name of his father and more distant ancestors, each prefixed by the term "ap," meaning "son of." The more the names, the more respected the family. There are remnants of the system in modern Welsh surnames such as Pugh (son of Hugh), Price (son of Rhys) and Parry (son of Harry). In Wales, Spain and almost everywhere else this practice is disappearing.

The telephone book in a reasonably settled part of the world (such as the mountainous country around Berne or, for that matter, in the isolated valleys of West Virginia) shows that different villages just a few miles apart each have a distinct set of surnames. In some villages almost everyone has the same one. Within each hamlet there has been an accidental loss of names as, by chance and over the years, some men have had no sons. Because the effect is random, different names have taken over in each place. The process may be helped by each village having been founded by a group which had, again by chance, its own characteristic set of surnames. It is not, of course, the case that within a village one name is somehow better than the others; its prevalence just reflects the accidents of history.

The genes of isolated populations, like their names, may also reflect a history of random change. Adjacent Yanomamo villages differ markedly in the frequency of blood groups and other inherited variants. Exactly the same is true in Alpine villages. Their blood group frequencies diverge to just the extent predicted from what their marriage records say about the number of inhabitants in each since they were founded. They have evolved by accident.

In modern Berne and in other cities, the picture is quite different. The phone book contains thousands of names, none of which is overwhelmingly common. Again, the rules of time and chance are at work. Cities contain so many people that it is unlikely that any name, or any gene, will go extinct simply because its few carriers fail to pass it on. Also, cities attract immigrants, so that new names (with their associated genes) come in all the time and the population becomes more diverse. A simple but effective way of measuring how genetically isolated a community might be is to count the number of names in relation to the number of people. If more or less everyone has a different name then the community is open to migration from many places. A glance at the New York telephone directory compared to that of, say, Oslo shows immediately that the two have had very different histories. The U.S.A. as a whole has a higher proportion of the global total of names than anywhere else. This reflects its background of immigration from all over the world.

Shared names usually mean shared ancestors which in turn means shared DNA. A population in which many people carry the same gene (or the same name) because they have inherited it from a common ancestor is said to be inbred. To some extent we must all be inbred as we are all to some degree related. Everyone has two parents, four grandparents and so on. If all had been independent and unrelated, the number of ancestors would double each generation. Taking a convenient landmark—1066, for example—and assuming that each generation takes twenty-five years arrives at a figure for the number of Britons at the time of the Norman Conquest of two to the thirty-seventh power, which is over a hundred billion. This, as the mathematicians say, is absurd. It shows that the lines of descent have merged and blended over the centuries. We all share many of our ancestors in common.

Perhaps the most inbred individual ever recorded was an aristocrat—Cleopatra-Berenike III, aunt of the Cleopatra who was enamored of Antony. She may have had identical copies of nearly half her genes because they descended from a single ancestor. As the ancient Egyptians saw the pharaohs as their gods' posterity they were anxious to keep the deities' bloodline as pure as possible by encouraging mating among relatives (sometimes, even, between brother and sister). The story is confused by difficulties in reading the hieroglyphs which show the degrees of pharaonic relatedness.

Levels of inbreeding vary greatly from place to place. The incidence of marriages between people with the same name is quite a good way of measuring it. This was first pointed out by George Darwin, son of the more famous Charles (who married his own cousin, Emma Wedgwood). George Darwin estimated from surnames that the proportion of cousin marriages (the closest legal form of inbreeding) among British aristocrats, by definition a small and exclusive group, was about four and a half percent. This was more than twice that in the general population of his time. The pattern of surnames shows that the British population as a whole is, on the average, more outbred than much of the rest of Europe. Even in remote and rural East Anglia, only one in fifty of the names present at the end of the eighteenth century had been there in the seventeenth, showing how much movement there had been compared to the situation in Switzerland or Italy.

In a small village, because there is not much choice when it comes to picking a spouse, relatives marry and the population becomes inbred. Sometimes the married couple have each received a copy of a harmful recessive gene from their common ancestor. As a result their children are at increased risk of having two copies of it. The consequences of increased inbreeding can be seen in the children of cousins. George Darwin picked up the effect. He found that Oxford and Cambridge oarsmen, presumably an unusually healthy group, were less likely to have issued from a cousin marriage than were their non-sporting peers.

There are, of course, constraints on how close a relative one may marry. Brother with sister is universally forbidden but even first cousin marriages may be illegal (as in most U.S. states in the nineteenth century and in Cyprus today). This social imperative may have arisen, in part at least, from a fear that the children might be less healthy. However, as childhood mortality was in any case so high when the taboos were formulated (so that a small increase because of genetic disease would not be noticed) perhaps they have no biological basis at all.

There certainly is an increase in the death rate (although by only one or two percent) and a slowing in the development of children born to close relatives. Cousins share a grandparent in common. If he or she carried a harmful recessive (as nearly everyone does) their children and grandchildren are more likely than average to inherit two copies. In some Japanese villages before the Second World War, up to a third of all marriages were between cousins. The huge survey of the population of Hiroshima after the atom bombs were dropped showed that the children of cousins walked and talked later than others and did worse in school. Part of this was due to the relative poverty of their parents but part reflects their genetic heritage. The same is true in India, where up to half of all marriages are still between cousins or between uncle and niece. The picture is confused here because such marriages tend to retain wealth within the family and to increase the number of children the parents can afford. Nevertheless, these too survive less well than the children of unrelated parents.

The effect can be seen in Britain. Nearly all the gypsies of South Wales belong to one extended kindred and nearly half their marriages are between relatives (which makes them one of the most inbred peoples on earth). One Welsh gypsy in four carries a single copy of the gene for phenylketonuria (see p. 184), which is four hundred times more frequent in this group than in Wales as a whole. A long history of inbreeding forced by social isolation has had an effect on their genetic health. First-generation Pakistani immigrants to Britain are also very inbred. Although only one birth in fifty is to such parents, about five percent of all inborn disease among British children is to those with Pakistani parents.

It is important not to overstate the dangers of inbreeding. Parents who are cousins have about a ninety percent chance of having a perfectly normal baby, compared to around ninety-four percent for unrelated parents. Inbreeding has an effect, but it is dwarfed by the improvements in child health which have taken place in the past few decades.

Partly because of inbreeding, isolated populations often show high frequencies of inherited abnormalities which are rare elsewhere. Among one small band of North American Indians, the Jemez of New Mexico, one person in a hundred and fifty is an albino. Finland has a legacy of nineteen different recessive genetic disorders each common in a particular community. The records kept by the Lutheran Church (to which ninety-eight of every hundred Finns once belonged) show that each local disease emerges from a history of marriage within the family when Finland was still sparsely populated.

Sometimes the effects of marriages of relatives are more subtle. A few women suffer from recurrent abortion. However often they become pregnant the fetus is lost. This problem is quite common among the Hutterites, a religious group who originated in the Tyrol in the sixteenth century and migrated to America in the 1870s. All thirty thousand of them descend from less than a hundred founders and marry only within the group. Over the years they have become very inbred. Hutterite women who find it difficult to have children share, it transpires, an unusually high fraction of their genes with their husbands. This is exactly what would happen if they were close relatives and may reflect the malign effects of inbreeding on the development of the embryo.

Sharing the genes which control antigens on the cell surface is particularly important. In lower animals, genetic variation on the surface of cells determines whether a sperm is allowed to fertilize a particular egg. If sperm and egg are too similar, then fertilization fails. Perhaps this is why the complicated system of genetic identification on the surface of cells evolved in the first place. The repeated failure of pregnancy in genetically similar husbands and wives may be the remnants of a method of ending pregnancies which arise from the attentions of too close a relative. Spontaneous abortion prevents such pregnancies from coming to term and producing a child who has two copies of a harmful recessive gene.

The same mechanism exists in more dramatic form in mice. Females can tell from scent how closely related a male might be. Given the chance, they avoid mating with their brothers. What is more, if a mouse pregnant by a relative is offered an unrelated male (or even the scent of his urine) she spontaneously aborts and mates with the new male. The genes responsible for mouse scent are closely linked to those which control cell-surface variation.

There is a hint of a similar mechanism in humans. Among the Hutterites, married couples are less likely to be similar to one another for certain genes in the immune system than are pairs who are just friends. The genes involved are related to those which drive sexual choices in mice. It seems that, quite unconsciously, most Hutterites—and, very probably, most people—fall for someone with a set of identity cues different from their own. Moreover, they are particularly keen to avoid a partner whose genes are too much like their mother's: the Hutterite mother (or perhaps anyone's mother) is to be avoided as a role model when choosing a wife. Just how it works, no one knows: but—if mice are a model—there is a suspicion that scent may be there somewhere.

Accidental genetic change is close to how God might play dice and the same statistical methods are used to study it. Population genetics is infested with mathematics, much of which is incomprehensible even to population geneticists. It is, however, unavoidable. The importance of random change depends on the size of the population. It is not enough just to know how many individuals there are around today. What is important is its average size since it began; after all, a large town may once have had just a few inhabitants. What is more, a special kind of average is needed. This pays particular attention to episodes of reduced numbers. Like so many ideas in evolution, the idea of the "harmonic mean" comes from economics. Think of a village in medieval times, with one rich squire and many starving peasants. Perhaps the fifty poor peasants each had an average income of a hundred guineas a year, while the squire gloried in a million. The average income was nineteen thousand guineas, which is a fairly meaningless statistic for anyone attempting to study the realities of ancient rural life. However, the harmonic mean income was a hundred and two guineas, which is a better reflection of what society was really like.

The same logic applies to populations that change in number. For example, the *average* size of a population whose size in succeeding generations was 1000, 1000, 10, 1000, and 1000 is 802 but its *harmonic mean* size is only 48. Any population bottleneck—ten individuals, in this case—has a dramatic effect which can persist for many generations.

There are other subtleties in trying to measure the real size of a population. If there is a lot of variation in the number of children produced by each person its effective size may be less than first appears. In many tribal populations (and perhaps in most ancient societies) there are big differences in reproductive success, particularly among males. A few Casanovas monopolize the females leaving lots of unwilling celibates who do not get their fair share. Freud, in *Totem and Taboo* (delightfully subtitled *Some Points of Agreement between the Mental Lives of Savages and Neurotics*), built his theory of psychoanalysis on this: a supposed time of a primal horde led by a dominant father with sexual rights to all the women. His sons killed and ate him, inheriting feelings of guilt and generating the Oedipus complex which has been such a nuisance ever since.

Many societies do have a rather Freudian structure (albeit without the cannibalism). In one Yanomamo village, four of the old men had 41, 42, 46 and 62 grandchildren respectively, while twenty-eight had only one grandchild and many more had none. Women, on the other hand, each had roughly the same number of descendants. Simply counting the men would greatly overestimate the real population size. From evolution's point of view, many of them might just as well not be there at all.

All this means that the uniquely male Y chromosome shows the importance of random change particularly well. Its inheritance is fairly close to that of surnames themselves, passing as it does from father to son. As only males have it, and as not all of them pass it on, the population of Y chromosomes is relatively small. There is little variation in its DNA sequence, perhaps because diversity has been lost through drift. In the Baruya tribe of Papua New Guinea all the men share one identical Y chromosome. Their common genetical surname may descend from a single male.

All populations have a history. The unyielding rules of chance mean that any episode of reduced size—a population bottleneck—will have a prolonged effect on later generations. From the very beginnings of antiquity humans have been colonizers, first as they filled the world from their African home and later as economic pressure drove people to conquer new lands. The emigrants were usually a small group, a tiny sample of the people left behind. The new colony may grow into millions, but all its inhabitants will carry only the genes (and the names) of the founders. As there were so few pioneers, the new population may be, purely by accident, quite different from those who stayed at home.

This "founder effect" is important throughout evolution. Darwin's first port of call on the *Beagle* voyage was the island of Madeira. He commented on how different its snails were from their European ancestors. This difference became even more striking when he began to look at the birds of the Galapagos. Perhaps, Darwin thought, the accidents of history, with chance colonizations of each island, explained why archipelagos were such natural laboratories for evolution.

The quirks of colonization have been just as important in the human past. Ironically enough, the best example of evolutionary accident comes from a return to Africa: the Afrikaners' journey back to their ancestral continent after an absence of more than a hundred thousand years. They began their migration from Europe in the 1650s. The pioneers brought with them a lasting legacy. It included more than Calvinism and bigotry. The surnames and the genes of their descendants are still a bequest from the first migrants. Although there are now two and a half million Afrikaners in South Africa they all derive from a small group of settlers, some of whom were so enthusiastic in their fecundity as to leave tens of thousands of descendants today. A million Afrikaners share just twenty names (Botha being one). This fits what history tells us about the number of immigrant families. Even today, half the most common surnames arrived before 1691 and the other half before 1717.

The migrants also brought, without knowing it, some rare genes drawn by chance from the people of Holland. One of the partners in the marriage of Gerrit Jansz and his wife Ariaantje Jacobs (who was one of a group of girls sent from a Rotterdam orphanage in the 1660s) must have carried a single copy of the gene for a form of porphyria. This disease (which is related to that which may have afflicted George III; see p. 73) is due to a failure in the synthesis of the red pigment of the blood. The symptoms vary from case to case. Sometimes, light-sensitive chemicals are laid down in the skin. Here, they react with sunlight and can produce painful sores. In some forms of porphyria hair grows on exposed areas. Sometimes the waste material accumulates in the brain, leading to mental disorder. Part is excreted in the urine, giving a characteristic port-wine, almost blood-red, color. Some claim that the origin of the werewolf legend—those creatures which come out only at night, howl and drink blood—may lie in the porphyria gene.

The South African form is relatively mild but became important when barbiturate drugs were used in the 1950s. Those with the gene suffered pain and delirium when they took them. Porphyria is rare in Europe, but thirty thousand Afrikaners carry it. In Johannesburg, there are more carriers than in the whole of Holland. All descend from a single member of the small founding population which grew in numbers to produce today's Afrikaners. Because it is so common in one family line, porphyria in South Africa is sometimes called "van Roojen Disease." A gene and a surname have become intimately associated and tell the same story about history.

The founder effect can be seen again and again among the descendants of those who colonized the world from a developing Europe. Sometimes, the settlements are isolated by miles of ocean. Tristan da Cunha, a tiny island in the South Atlantic, was settled in by a garrison sent to guard Napoleon, then in exile on St. Helena. A few soldiers stayed on after the guard was withdrawn. They obtained wives by advertising, and a few shipwrecked sailors and others joined the community over the years. It went through a second bottleneck when several men drowned in a fishing accident and some families moved away, heeding the advice of a gloomy pastor. Now, there are a few hundred people on Tristan. Again, they share names—Bentley, Glass and Swain, the names of three of the first founders, are still common— and there is a characteristic local genetic abnormality, a hereditary blindness brought by one of the original wives.

Some migrant communities are isolated by social rather than physical barriers. A tiny group of Jews have lived in Kurdistan since the Diaspora, nearly three millennia ago. They kept themselves rigidly separate from the surrounding Islamic peoples. The Kurdish Jews (who have now returned to Israel) show an extraordinarily high frequency of an inherited enzyme defect. In the Middle East as a whole about one person in fifty carries this gene, but two thirds of the Kurdish Jews have it, presumably because one of their remote ancestors was a carrier. Even in the United States there are many religious groups whose founders emigrated to avoid persecution. They have grown into large populations which rigidly exclude outsiders. The Pennsylvania Amish, stars of the film *Witness*, have a unique inheritance. Nearly a hundred babies have been born with six fingers and restricted growth, a condition almost unknown elsewhere. Every one of the affected children descends from Samuel King, a founder of the community.

Tracing the movement of a gene around the world also illustrates the importance of chance in evolution. Huntington's Disease (see p. 69) is relatively common among Afrikaners. Most of the cases descend from a Dutch man or his wife who emigrated in the 1650s. All copies on Mauritius are the legacy of a French nobleman's grandson, Pierre Dagnet d'Assigné de Bourbon, and more than four hundred patients in Australia have inherited their gene from a British immigrant, Mrs. Cundick. In Wales, there is a patch of Huntington's Disease in the Sirhowy Valley, around the house of a mason who settled there in the nineteenth century and who must have carried it. The largest kindred of the world (which is being used to track down just where the gene is located) is in Venezuela around an arm of the sea called Lake Maracaibo. Nearly ten thousand living descendants of one Maria Concepción, who died in about 1800, have been traced. Four thousand of them either have the disease or are at risk of contracting it.

This sort of thing must have happened again and again in earlier times as humans spread across the world. Even if written history did not exist, the surnames of the Afrikaners would allow a pretty good guess to be made about how many people were in at the beginning, three hundred and more years ago. Genes can be used to do the same job. Patterns of variation show how many people founded a population, or whether it went through a bottleneck in the distant past.

There are some striking global patterns in the extent of inherited diversity. Africans are more variable than are the rest of the world's peoples. Their cell-surface antigens—the inherited cues recognized by the immune system—show about twice as much variation as do the equivalent genes in Europe, and many antigens are unique to Africa. Africans are more variable for blood groups, proteins and DNA sequences as well. For mitochondrial DNA, the average difference between two Africans is twice that found elsewhere. Venezuelan Indians, whose ancestors were near the end of the long history of movement across the world from Africa, have no variation in their mitochondrial DNA at all.

The decrease in diversity outside Africa, humankind's native continent, may be because many genes were lost as small bands of people moved, split and founded new colonies during the long trek across the globe. Just as for the Afrikaner surnames the number of variants dropped each time a new colony was founded. The high levels of diversity among Africans is evidence that *Homo sapiens* has been evolving in Africa longer than anywhere else. Its decrease at the tips of the evolutionary branches in South America and Polynesia suggests that much of human evolution was driven by chance as the migrating population passed through a succession of bottlenecks.

Comparing the genes of Africans with those of their descendants elsewhere in the world even makes it possible to guess at the size of some of these early colonizations. The order of bases along a small piece of DNA is in some ways a "genetic surname," a set of inherited letters which pass together as a group down the generations. Only one DNA segment of this kind, that around one of the hemoglobin genes, has been looked at in much detail worldwide. The results of the survey are startling.

All populations outside Africa, from Britain to Tahiti, share a few common sequences for this section of the DNA. Within Africa, there is a different pattern of distribution. Just like the names in the Johannesburg telephone book compared to that of Amsterdam, the shift in pattern from the ancestral continent to its descendants may be a relic of a population bottleneck at the time of migration—this time from, rather than to, Africa. We can do some statistics (and make quite a lot of guesses) to work out the size of this hundred-thousand-year-old event. They show that the whole of the world's population outside Africa may descend from a group of less than a hundred emigrants. If this is true, non-Africans were once an endangered species.

Because of the mating habits of males, with many forced into involuntary celibacy, Y chromosomes are particularly susceptible to the effects of time and chance. The Y chromosomes of Europe and Asia fall into just two genetically distinct groups. Perhaps all modern European and Asiatic men can trace their ancestry to one of two ancestral males—Adams, as it were. Unfortunately, too little is known about the Y chromosomes of Africans to estimate how many pioneering men there might have been.

There are two cultures in science: one (to which most scientists belong) uses mathematics and the other understands it. All these guesses about ancient population bottlenecks demand some daring statistical acrobatics. Sometimes these have proved too much for those who indulge in them. The much-publicized idea that women can be traced back through the mitochondrial DNA to a single female—African Eve—who lived two hundred thousand years ago has suffered this fate. Some statisticians claim that there is simply not information available on the world's mitochondria to trace a common ancestor and—although those who made the initial discovery dispute the suggestion—Eve may be (at least temporarily) dead. All the guesses (including that of the number of African emigrants) also depend on one crucial, and perhaps quite mistaken, assumption; that the genes involved do not alter the chances of their carriers surviving and reproducing. Molecular biologists tend to assume that small changes in the structure of DNA are unimportant. It is equally conceivable that they do have an effect on survival. If, for example, Africans have more variation on the surfaces of their cells because it helps to combat disease, then to claim that a reduction elsewhere is due to a bottleneck is simply wrong.

Any attempt to reconstruct the distant past is bound to suffer from ambiguities such as these. Although genetics has not yet revealed just how many Adams and Eves there were, it is clear that much of the human condition has been shaped by accident. This might at least instill a certain humility into those whose genes have defeated the laws of chance by surviving to the present day.

8 | The Economics of Eden

RENAISSANCE PAINTERS on religious themes had a problem: when they showed Adam and Eve, should they have navels? If they did, then surely it was blasphemous as it implied that they must have had a mother. If they did not, then it looked silly. Although some compromised with a strategic piece of shrubbery, this did not really solve things. And where was the Garden of Eden? Various theories had it in Israel, Africa and even the United States. *When* it existed seemed obvious; adding up the ages of the descendants of the primal couple given in the Bible set the beginning of history as 23 October 4004 B.C.

The reason for leaving Eden was also clear. Its inhabitants had, with the help of an apple, learned something new, and as a punishment were forced out into the world. No longer could they depend on a god-given supply of food falling into their hands. Instead, they had to make a living for themselves. The first economy was born.

This chapter is about the escape from Eden: about colonizing the world and how genetic change is linked to the earliest economic developments. Economics is often seen as a kind of enlightened self-interest. The desire to increase one's own wealth may, as Adam Smith has it, be the invisible hand which is at the foundation of all social progress. The same argument has been used by some evolutionists. Genes are seen as "selfish" and anxious to promote their own interests, even at the expense of their carriers. In its most naive form, this view of life is used to explain (or at least to excuse) spite, sexism, nationalism, racism and the economic and political systems which grow from them.

There are some obvious ties between theories of economics and of evolution. Darwin was much influenced by the works of the early economist Malthus, who had been disturbed by the new slums of the English cities of the eighteenth century. In his *Essay on the Principles of Population* Malthus argued that populations will inevitably outgrow resources. Darwin in his autobiography wrote that it was reading this that first gave him the idea of natural selection.

Karl Marx, himself a denizen of one of the most congested of all London districts, was equally impressed by the dismal conditions of the new proletariat. He sent Darwin a copy of *Das Kapital* (which was found unread after his death). Marx, in a letter to Engels three years after *The Origin of Species* was published, went so far as to say that "It is remarkable how Darwin recognises among beasts and plants his English society, with its division of labour, competition, opening up of new markets, inventions, and the Malthusian struggle for existence." Engels took it further. In his essay *The Part Played by Labour in the Transition from Ape to Man* he argued that an economic change, the use of hands to make things, was crucial to the origin of humans. If one substitutes the term "tools" for "labour" his views sound remarkably like those of modern paleontologists.

Genetics shows that much of evolution is, as Engels said, linked to social advance. However, far from society being impelled by its genes it seems that social and economic changes have produced many of the genetic patterns seen in the world today. Since modern humans first appeared every technical advance has led to an evolutionary shift and to biological consequences which can persist for thousands of years. Society—particularly the economic pressures which lead people to move—drives genes, rather than genes driving society. Our relentless expansion is at the center of our evolution. In Pascal's more pessimistic words, "All human troubles arise from an unwillingness to stay where we were born."

Fossils show that almost as soon as they evolved, humans began to migrate. Why our ancestors were so restless, nobody knows. Technical progress may have had something to do with it. Although the emergence of modern humans did coincide with improvements in the making of stone axes and the like, tools had been made for at least two million years before the great diaspora which filled the world.

Perhaps climatic change was involved. The Sahara Desert was once a grassy plain and Lake Chad a sea bigger than the present Caspian. Both dried up about a hundred thousand years ago, so that food shortage (which meant that populations outgrew resources) may have driven the first humans out of Africa. A microcosm of what happened then is taking place now at the southern edge of the Sahara. As the rains fail, the desert is spreading into the Sahel and migrants are on the move.

The earliest economies were based on exploiting the animals and plants which were available. They had simple foundations. People used what nature provided, until it ran out. All over the world there are fossils of large and tasty animals which were driven to extinction soon after humans arrived. In Siberia, so many mammoths were killed that the hunters made houses from their bones. In Australia, too, there was a shift from forests to grasslands as the first immigrants burned their way across the continent. The record of destruction is preserved in the Greenland ice sheet. The snows which fell tens of thousands of years ago retain the soot and ash from gigantic forest fires, some of which may have been set by human beings.

New Zealand was not colonized until about the time of William the Conqueror. For a few years there flourished a culture based on the exploitation of a dozen species of moas, giant flightless birds. Relics made of their feathers are still around, as are the ritual slaughtering grounds where the birds were killed (and where half a million skeletons have been found). Not surprisingly, the birds themselves were driven to extinction within a few centuries. Even in Europe whole faunas went quite recently. Humans did not reach Crete, Cyprus and Corsica until around ten thousand years ago. Before then they had some extraordinary inhabitants: pygmy hippos, deer and elephants, and giant dormice, owls and tortoises. Soon after the arrival of the first tourists, all were gone, and the burnt bones of barbecued hippos are scattered among the remnants of the earliest Cypriots. Perhaps the Greek legend of Cyclops—the monster with a single huge eye in the center of his forehead—originated in the skull of a dwarf elephant, with its central aperture for the trunk, coming to the surface after many years.

The common large mammal in Europe and the Near East at the time when modern humans were moving eastward from Africa on their journey to Australia was one of their own relatives, Neanderthal Man. He had lived there quite happily for two hundred thousand years. There were many Neanderthals in the dense forests of southern France. Some had an economy based on hunting reindeer, with settlements concentrated around their migration routes. In the cave of

Combe Grenal in Périgord tens of thousands of Neanderthal stone tools of more than sixty different types have been found. Neanderthals preserved food by burying it in the frozen ground. In several places, there are what may be Neanderthal graves. Sometimes burials were formal. A fifty-thousand-year-old Neanderthal grave at Teshik-Tash in Uzbekistan is surrounded by ibex horns, perhaps as a religious ritual.

Neanderthal culture was, in its own way, a sophisticated one. However, it was remarkably unprogressive, with no real change for a hundred thousand years. Tools made by Neanderthals in Britain and the Middle East look almost the same. They had little interest in exploration and never made boats, so that the delights of the Mediterranean islands (hippo-infested though they were) remained unknown. Neanderthals were the first conservatives.

Fairly soon after the invasion of Europe by our own direct ancestors, they disappeared. Why, we can only guess. The guesses range from genocide to interbreeding. The first is unlikely. In France, at the cave of St. Césaire, Neanderthals and moderns lived close to each other for thousands of years. The second is probably wrong. If there had been extensive mating between the indigenous population and the invaders, then modern Europeans would be expected to retain genes from this distinct branch of the human family and to be genetically different from, say, today's Chinese or Indians, whose ancestors never met a Neanderthal, let alone mated with one. They are not.

Perhaps economic pressure did the job. For most of human history, Africa was the most economically advanced of all the continents. Africans were making sharp stone blades when Europeans had to make do with blunt axes. There was a period when Neanderthals seemed to pick up some of the new technology, but this did not last for long. The first non-Neanderthal Europeans were found in 1868 during railway work at the Périgord village of Les Eyzies, in the Cro-Magnon shelter. Cro-Magnons looked much like modern Europeans. They (and their immediate predecessors the Aurignacians) had a sophisticated hunter-gatherer economy with a variety of tools. Their cave art reached its peak around forty thousand years ago. The moderns had tools made of bone and ivory when their Neanderthal relatives still had to make do with stone. They were better at exploiting what was available, so that their populations grew faster, driving Neanderthals (and their genes) out. The last known Neanderthal skeletons are from St. Césaire. They died more than thirty thousand years ago.

Simple though it was, the Neanderthal economy kept modern humans at bay for a long time. The moderns reached Australia before they filled Europe. Competition from its indigenous Neanderthals may have had something to do with the delay.

Most of the globe was populated quite quickly after humans left their natal continent. The first Australians arrived early on, about fifty thousand years ago. The earliest remains are in two sites in Arnhem Land, in north Australia, where there are stone tools and red or yellow ocher paints in a sandy deposit of that era. The sites are close to the shore and perhaps to the point where humans arrived from the north. The oldest known Australian skulls date from thirty thousand years ago. Within five thousand years the ancient Australians had complex tools and fishing nets and were economically as advanced as the rest of the world's peoples.

For much of its history Australia was joined to what is now New Guinea by a land bridge which disappeared only seven thousand years ago. Tasmania was also part of Greater Australia. This continent, Sahul, has always been separated from Asia by a deep trench. Its existence was guessed at by Darwin's collaborator Alfred Russel Wallace who noticed a great change in the animals and plants in that part of the world. The first Australians must have crossed at least fifty miles of water to reach their new home.

Difficult although this passage may have been, the genetics of today's native Australian populations suggest that it was made by many people. The mitochondrial DNA of Australian aboriginals and Papuans is quite diverse. There must have been many founders, with several incursions into the continent. Once they got there, the new inhabitants seemed to find the habitat congenial. At least in the tropical north they tended to stay put. In Papua New Guinea local populations are genetically very different from each other and there are distinct "clans" of mitochondrial lineages, each limited to one or two remote mountain valleys. Their denizens have been isolated for a long time—and remained so until the first Europeans reached the interior half a century or so ago. The early Papuans were in their own way economically advanced, cutting down trees to allow the tastier plants beneath them to grow. By staying in their isolated fastnesses for tens of thousands of years they remained insulated from the economic strife and the waves of population movement which affected the rest of the world.

At the other end of Sahul rising sea levels soon marooned the inhabitants of Tasmania. The Tasmanians knew nothing of the cultural progress of the mainland and remained ignorant of the outside world until the arrival of the next wave of immigrants, the Europeans, in the eighteenth century. Nothing is known of the Tasmanians' genes, for a simple reason. They were driven to extinction (and sometimes hunted down) by the representatives of the modern world's economy within a few score years. There was a sordid episode in anthropology when the Tasmanians were regarded—absurdly—as the elusive "missing link" between humans and apes and the museums of the world quarreled over the bones of the last survivors. Now there is a new chance to look at their inheritance, as some of the prehistoric art of Tasmania includes bloody handprints and perhaps ancient DNA.

Human traces on Pacific islands show that even remote islands (such as Manus Island in the Admiralty group, over two hundred miles from the nearest land mass, New Ireland) were occupied twenty-eight thousand years ago, so that by then it was possible to make substantial voyages. The genes of present-day Melanesians, those inhabiting the islands north and east of the main Australian land mass, are still fairly similar to those of the ancient populations in the Papuan highlands. They are the descendants of these early voyagers.

The Polynesians who occupy the rest of the Pacific including the remotest islands such as Hawaii are quite different and got there much more recently. Hawaii and Easter Island were reached only a couple of centuries after the birth of Christ. In the far Pacific, islands separated by thousands of miles of ocean are not genetically very distinct, showing that water is a less effective barrier to movement than is land.

Nearly all the peoples of the distant Pacific carry a small change in their mitochondrial DNA. A nine-letter part of the genetic message is missing. This deletion, as it is known, has spread through the whole of Polynesia from Fiji to New Zealand. In some places it is so common as to suggest that most of the present population descends from a single founding female who was the ancestor of nearly all the inhabitants of the Pacific islands. The deletion is shared with the populations of east Asia, such as the Taiwanese and the Japanese. It seems that the Polynesians spread across the Pacific from Asia and not from Australia. Australian aboriginals and the highlanders of Papua New Guinea do not have this unique genetic signature. This confirms the evidence from archaeology that they are the descendants of an earlier migration from Africa, which began thousands of years before that of the Polynesian *arrivistes*.

One thing is clear: there are few genetic links between the peoples of the Pacific and those of South America. Thor Heyerdahl's book of his daring voyage in a balsa raft across eight thousand miles of Pacific from Peru has sold twenty million copies, probably more than all other anthropology books put together. Unfortunately, his view that to reconstruct the past it is only necessary to re-enact it is wrong. Population genetics has sunk the *Kon-Tiki*.

Twenty thousand years ago, much of the Pacific had a dense population and a prosperous economy. In Europe, too, trade was well advanced. Flint for stone tools was transported for many miles and Baltic amber reached the Mediterranean. There was a brief flourishing of art, perhaps only a couple of centuries long. During that time, the caves at Lascaux and Altamira were filled with images, and statuettes and shell necklaces began to appear.

While the world economy boomed the Americas were empty. They were finally reached from Siberia. Many of the inhabitants of that icy land, which was even colder than it is today, lived by hunting mammoths. As they spread they destroyed their food sources. At last, they came to the Bering Land Bridge which joined Asia to Alaska. It emerged from the sea—as did thousands of square miles of coastal plains all over the world—as water was locked into the ice. At the end of the ice age the water rose and twelve thousand years ago the bridge between Old and New Worlds was breached. Just before it disappeared, a few pioneers made their way across. If their experiences were like those of what we know of the nineteenth-century Inuit who made long voyages across similarly barren landscapes they had a grim time. Many must have starved. Nevertheless, some reached the broad plains of North America and quickly spread to the continent's southern point, reaching it within only a couple of thousand years. This seems like a rapid expansion but is, after all, less than ten miles a year into a deserted landscape. The journey was helped by a brief warming which meant that, even in Alaska, a few trees appeared in the bitter landscape.

Once again, the edible inhabitants suffered. Mammoths, sloths, giant tapirs and camels followed each other into extinction. Each was large, tasty, naive and tame. They reproduced slowly. Once humans had arrived their fate was certain. Perhaps the wave of destruction of resources is what tempted the first Americans southward until, in Patagonia, they could go no farther. The evidence of devastation persists into historical times. The Cedars of Lebanon were an image of wealth, used by Solomon to build the temple. Now, there are just a few left. In Spain, the Mesta, the great cooperative of the shepherds, succeeded in turning most of the country into a desert within three hundred years. And no one needs to be reminded of the latest attack on the ecology of the Americas as the rainforest burns.

The date of the American invasion is not certain. The oldest traces of occupation in North America are in a rock shelter in Pennsylvania. They date from about twelve thousand years ago. Soon, the "Clovis culture," in what is now the United States, was producing sharp and effective arrowheads. The first art in the Americas is at the cave of Pedra Furada—the Perforated Rock—in Brazil, where there are twelve-thousand-year-old paintings of birds, deer and armadillos, together with human stick figures. There are claims that charcoal from nearby caves dates back for fifty thousand years, but few anthropologists accept this as evidence of human occupation. Most believe that the first Americans arrived less than twenty-five thousand years ago.

The genes of Native Americans fit well with the idea of a small founding band from Siberia who quickly filled their new-found land. Americans as a whole are much less diverse and more geographically uniform than are the peoples of highland Papua New Guinea (who fill only a tiny proportion of the space occupied by those of the New World). The mitochondrial genes of all Native Americans fall into just four distinct lineages suggesting that just a few people managed to complete the hazardous crossing of the Bering Bridge. The same lineages are found in some three-thousand-year-old Chilean mummies, implying that there were not many bottlenecks on the way through the Americas from north to south. Although nothing is known of the genetics of today's Siberians, the mitochondria of South American Indians are generally similar to those of northeast Asia, supporting the idea that their ancestors came from there.

By ten thousand years before the present humans had filled the whole habitable world, apart from some remote islands. Everywhere they lived in small bands. Every Englishman needed ten square miles of land to feed himself. The global spread was accompanied by technical advances in axes, arrowheads and fishing nets as the most easily exploited animals—reindeer, mammoths, giant kangaroos or emus—disappeared and the hunters were forced to move to less easy prey.

Studies of the genes of the few modern peoples who still depend on hunting and gathering to fill some of their needs say something about the way of life of our hunting ancestors. Adjacent villages among the Yanomamo Indians differ quite markedly from each other in enzymes and blood groups, evidence that their social structure, based as it is on suspicion and hatred, has led to genetic isolation. There were further opportunities for random genetic change as each hunting band split and moved on during the peopling of the globe. No doubt the life of a hunter-gatherer was a lonely one. Although the immediate group may have been closely knit, there was little contact with anyone else.

Ten thousand years ago, everything changed. There was an economic breakthrough that was to shape the society and the genes of the modern world. Farming appeared.

Before agriculture, humans ate dozens of kinds of food. An excavation in Syria shows that there were then more than a hundred and fifty species of food plant, but after farming began only a few cereals and pulses. Even in the nineteenth century, North Queensland aborigines ate two hundred and forty different species of plant. Adding together the top five crop plants for every country in the world today gives a global total of only a hundred and thirty species. Some foods which are now staple items were absent from farmers' diets until quite recently. Chickens were imported to Europe from India long after sheep and cattle had been domesticated. In ancient Greece there were no eggs for breakfast.

Hunters had an easier life than the first farmers. The modern !Kung Bushmen who still live in this way only need to work for fifteen hours a week to feed their families, far fewer than those who have moved to the farming economy and less than the time which most western industrial workers have to spend at work to buy food. In the Middle East, too, wild grasses growing on hillsides are abundant enough to allow a family armed with primitive sickles to gather enough seeds in a few weeks to feed themselves for a year. Perhaps the extra work involved explains the Bible's disparaging tone about the new economic system: Adam on the expulsion from his hunter-gathering Eden was admonished, "Cursed is the ground because of you; in toil you shall eat of it all your life . . . therefore the Lord God sent him forth from the Garden of Eden to till the ground from which he was taken."

The first of all farmers lived in the Middle East, probably in the basin of the River Jordan (which is, incidentally, close to where the Biblical Eden must have been). There was plenty of natural food around the Lake of Jordan. It was difficult for the people who lived there to move elsewhere when times got bad, because of the surrounding deserts. Ten millennia ago the weather began to change. There had been a "continental" climate rather like that of the Midwest of the United States today. Winters were cold and wet and although the summer was hot there was plenty of rain. Suddenly it shifted toward a Mediterranean climate with warm wet winters and hot dry summers. The lake began to dry up, and what was once a continuous sheet of fresh water split into the salty Sea of Galilee and Dead Sea.

Pollen shows that the plants began to change too. The forests shrank and grasses took over. Mediterranean climates are very good at fostering the evolution of new plant species. California, the tip of South Africa and Western Australia are all Mediterranean in climate and each is the source of hundreds of unique plants. In the Jordan valley there were new and fertile hybrids between grass species which came together as the countryside dried. The local people burnt the grass to attract deer and gazelles to its new shoots. Soon they had the idea of planting the seeds; and farming began. The teeth of the ancient inhabitants of the valley are worn, because the first grains were milled on soft grindstones and their food was full of grit.

The same sort of thing happened at roughly the same time in several places. In each there was a transition period which involved burning grass to harvest the new shoots and even irrigating wild stands of vegetation. Agriculture spread quickly. Wheat was first cultivated in the Middle East, rice in China and maize in South America. Somewhat later there was a domestication of sorghum, millet and yams in West Africa. The effect was always the same. There was a population explosion. Before farming, each person needed about a square mile to feed himself. After it, a hundred people could live off the same space.

Fossil bones suggest that the health of farmers, far from improving, actually got worse. Deficiency diseases appeared as the amount of protein in the diet went down and there were periods of starvation as population outgrew resources. In some places the effect was striking. If children eat well, they grow up tall. This is why the average height in most Western countries has gone up by three inches during the past century. For the children of the first farmers—like those of the proletariat of the Industrial Revolution—exactly the opposite happened. In southeast Europe the average height of men fell by seven inches in the millennium when farming began. The bones of North Americans show extensive damage, particularly in the eye sockets, as maize became the main foodstuff. Maize has very little iron and protein and, even worse, reduces the absorption of iron from other sources such as meat. This led to an outbreak of anemia, whose record is preserved in the skulls of those who depended on the new maize economy.

Population growth meant that farming spread rapidly from its centers of origin. Waves of technical change radiated from each of them. In Europe, decorated beakers appear in archaeological digs, and in the Far East implements of rice cultivation spread for thousands of miles from their Chinese homeland.

European farming began in the Middle East about ten thousand years ago. It reached Greece about five thousand B.C. and took more than two millennia to cross Europe from there. Its expansion was not regular. The moving agricultural frontier was rather like that of the nineteenth-century Wild West. The colonists settled the best areas first, leaving the less valuable lands to their original inhabitants. In north and east Europe, hunter-gatherers were doing well enough to stall the wave of farmers from the Danube basin for a thousand years. The northward spread of farming was further slowed by a worsening climate which made it hard to grow crops. The new technology did not reach the North Sea coast until about five thousand years ago. It quickly spread to Britain from there. Elsewhere, it was delayed for even longer. In southern Finland a farming economy did not begin until after the time of Christ.

Much of the resistance to the new way of life was due to the success of the local hunting economy, the "Forest Neolithic." Nine thousand years ago northern Europe had a population of affluent foragers. They lived in large camps, built traps for their prey, and stored great caches of food. Around the Baltic, they built stilt villages in ice-dammed lakes. In some places, hunters specialized in seals and in others in deer. Those who did the gathering ate thirty or more different plants— grasses, acorns, sorrel and dandelions and, in marshy places, water chestnuts. Millions of broken water chestnut shells have been found, together with the wooden mallets used to smash them. The only crop was flax which was used for rope rather than food.

Wherever farming arrived, the local hunter-gatherers suffered, sooner or later, a process of gentrification—or even yuppification—as a wave of economically advanced people moved in on them. It is easy to imagine the complaints of the natives as the newcomers with their new-fangled ways and high technology imports disrupted their rural idyll.

Although the farmers did finally overwhelm the hunters there was a long period of coexistence. Preserved relics show that the farmers traded grain for meat and furs. In some places, the transition from the old to the new economy took several thousand years, with a slow decline in the number of bones of wild pigs and deer and of natural grasses (as shown by the impressions of their seeds in fragments of pottery) in favor of cattle and grains. The worsening climate finally ended hunting. Oysters and seals disappeared from the Baltic and the northern hunters at last moved into the modern world.

Economic historians have two views of the origin of technology. One theory, the diffusionist, has it that all technical advances pass by learning from community to community. Knowledge itself moves, rather than the people who know. The other claims that cultural advance comes from displacement and the conquest of one people by another. The culturally advanced bring their knowledge with them and replace their predecessors. There are many hints about the economy of the European Community ten thousand years ago in bones, pots and seeds; but the genes tell us more. Genetic patterns in today's Europeans show that both migration and diffusion were involved in the replacement of hunting with agriculture. The farmers did move in on the hunters, but social barriers do not seem to have stopped people mating across the class divide.

A genetic map based on two dozen variable genes from three thousand places in modern Europe shows some clear trends. Most are from southeast to northwest, from Greece to Ireland. This map looks very like one of the wave of advance of farming made by charting the spread of agricultural implements and the like in archaeological digs. Farmers advanced at about half a mile per year, probably by founding new farms at the edge of their expanding population. It seems that they interbred with the local hunters and, because the farmers were so much more numerous, absorbed their genes. This process began in the Balkans and was completed thousands of years later on the western fringes of Europe to give the genetic patterns seen today. By the time the farmers reached the far north and west their genes had been much diluted by mixing with those of the aboriginal Europeans. The British contain more hunting genes than do, say, the Greeks, who descend from a less adulterated wave of farmers who had rolled over the earlier economy and absorbed its genes long before. The biological heritage of hunters and farmers means that today's British are more closely related to the Portuguese than to the Yugoslavs. The latter live about the same distance away but are closer to the Middle-Eastern source of farming.

The genetic map of Europe has a few striking anomalies. The Basques do not fit at all into the general pattern. They have a number of unique features—for example, they have the highest frequency of the Rhesus negative blood group gene in the world. Excavations in the Basque country show that the locals resisted the new farming technology for thousands of years. They still differ from all other Europeans. Basques may be closer to our hunting ancestors than anyone else. The Lapps, too, are quite distinct and seem to descend from a different group of hunters—whose way of life they still in part retain. Sardinians are rather different from the rest of Europe and have affinities with the Basques. Their island home may have reduced the number of farming immigrants.

Although less information is available there also seem to be genetic trends away from a Middle-Eastern center to the northeast toward Siberia, the southeast in the direction of India, and southwest into North Africa. Perhaps these too are a reflection of a wave of farmers moving away from their booming populations in every direction and absorbing the genes of the local inhabitants as they spread.

The farmers left genetic trails in other parts of the world, too. Rice cultivation started in the Yangtse basin about eight thousand years ago. Within three thousand years there were rice farmers from Vietnam to Thailand and north India. These were the people who developed sea-going canoes and spread into the remote Pacific, where—because rice cannot be grown there—they planted breadfruit, taro and yams. The pollen record from three millennia ago shows that large parts of Java were intensively farmed. Because they were moving into an empty land the genes of these Pacific farmers and fishermen today are still quite similar to those of their Asian ancestors. In Africa, too, there was a population explosion in places where millet was first grown. This has left genetic trails across Africa. The movements of the sickle-cell gene can be traced across the continent in the wake of the first farmers.

These early African farmers and their counterparts in the rest of the world no doubt experienced social unrest as they gave up hunting to move to a more productive but perhaps less enjoyable way of life. However, any romantic view of a socially harmonious past when contented foragers shared everything is simply a longing for a non-existent Golden Age. Virgil, in the *Georgics*, mourns for a time when "No fences parted fields, nor marks nor bounds,/ Divided acres of litigious grounds." His plaints over a happier past may have been shared by the early farmers regretting the departure of the glorious times when they hunted food rather than grew it. Whatever the truth, the beginning of agriculture marked the end of an economic system based mainly on individual effort which had persisted for nine tenths of history. With farming, Eden had been left forever; and politics began.

9 | The Kingdoms of Cain

ADAM AND EVE'S CHILDREN were a worry to their parents. Their eldest, Cain, is best known for having killed his brother Abel. He has another distinction as well. Just one generation after the expulsion from Eden he became the primal capitalist. As the Old Testament says, he was the earliest to "set bounds to fields." By so doing he erected the first barriers among the peoples of the world. Frontiers have driven society, history, and genes ever since.

No doubt the idea which came to Cain struck the early farmers as well. With agriculture, the idea of ownership of land was born. The process can be seen today as hunter-gatherers give up their old way of life. The Kipsigis of Kenya moved to a settled existence as maize farmers early in the present century. Great inequalities of wealth soon appeared, depending on who obtained the best land. When harvests were bad the poor starved while the rich grew fat. Competition among males to gain a mate increased, so that the beginning of farming marked a new campaign in the battle of the sexes. Those with productive land had far more children than did those without. Perhaps the same social dislocation affected farmers everywhere. The genesis of agriculture may have been when social class really began. From Mycenae to ancient Chile there was—as there is today—a difference in height and health (as manifest in bone damage in disinterred skeletons) between the rich, interred with their ornaments, and the rural poor, buried in poverty.

No doubt the conflicts among the first farmers extended to arguments about who grew what and where. It did not take long for property to pass into fewer hands and for society to evolve toward the system of competing nations which we see today. Any barrier, be it a mountain, a frontier, or an inability to understand, which stops peoples from meeting and mating will cause them to diverge. All over the world today there are genetic changes which mark the divisions—the bounds to fields—between ancient societies.

Barriers based on politics are new things in evolution. Genetics suggests that what we see as history, the struggles between peoples, is a recent event. From the Old Testament to *Mein Kampf*, historians have seen mass movement and conquest as the key to the peoples of the world. In the turbulent years after the First World War, the League of Nations tried to define just what a "nation" actually was. Depressingly enough, the best they could come up with "a society possessing the means of making war." Over the past half millennium, most great states have spent half their time at war. However, marauding nations began to shape genetic history only in the past few thousand years. Before then, people and their genes moved by gradual diffusion or by migration into an empty land, rather than by the defeat of one country by another.

In many parts of the world—including the Jordan valley—the first farms were by rivers in an arid landscape. Such rivers (the Nile most of all) often flood, leaving fertile silt as they recede. Modern tribal farmers who use the land left bare by the departing waters of the Senegal River get a return on their labor of fifteen thousand percent: for every calorie of effort they put in they get a hundred and fifty back as food. This compares with a return of around fifty to one for the most efficient irrigated rice fields. The return on the flood plain is enormous but, like the Hong Kong stock market, unpredictable. Some years are excellent but others are dry and disastrous. The flooding of the Nile has been recorded from AD 641 to today. There has been a hundredfold difference in the area of land submerged from year to year.

In today's Senegal this has produced a rigid pecking order. Some families always have access to the floodlands, even when the area inundated is small. Others are allowed to grow crops only when the river has risen high and covered large tracts of ground. In dry years they have to find food elsewhere: in earlier times, probably by a return to hunting and gathering. Perhaps the earliest settled communities developed, not to increase the efficiency of farming, but to manage risk. A wild free-for-all for the best land in a bad year would have been dangerous and expensive; society evolved as a way of coping with uncertainty.

Ten thousand years ago the Natufians, the descendants of the first farmers in the Jordan valley, had built villages with timber houses. Within two millennia, and perhaps before, there were much larger villages in Mesopotamia. It took only a few centuries for civilization to advance to such an extent that the settlements were surrounded by walls, ditches and watchtowers. Warfare had begun to play the part which it has retained ever since. Farmers were forced from their earliest villages by land degradation and the pressure of numbers. In Mesopotamia they moved into the hot and dry plains some distance away from the Tigris and Euphrates rivers. Soon, the earliest city-states began, perhaps because of the need to organize which began with the invention of irrigation. For the first time humanity was divided by political rather than physical barriers. Genetic patterns in the modern world show that since that day bigotry has sometimes been as effective an obstacle to the flow of genes as has geography.

The first capitalists, like those of today, were helped by technology. Six-thousand-year-old horses at Sredny Stog in the Ukraine have broken teeth, suggesting that they were controlled by bits. Riding increased mobility and helped people to work together to steal resources from others. The power of the horse in conquest is seen in the success of a few dozen Spaniards in destroying the Inca and Aztec Empires and of the Mongols in taking over Hungary. Soon after the appearance of horsemen the civilizations of Eastern Europe built defensive walls around their towns. Within a few years their societies had collapsed.

By 3600 BC there were great cities in Mesopotamia. Uruk had ten thousand people and within a millennium that number had increased fivefold. Its growth was due in part to warfare. Scores of villages were abandoned by their people, who moved to the new cities. The Sumerian city-states, the first organized political entities in history, were the source of writing and of wheeled transport. They had a priestly and an aristocratic caste; and, of course, a dispossessed mass. Their decline was hastened by mismanagement. As irrigation continued, the soil became salty and in the last years of Sumeria crop production dropped to a third of its peak. Finally these, the first national units, were overcome by one of the first empires, that of the Akkadians, who invaded from the north.

Other early cities also came to an end because of bad planning. The deserted city of Petra, in Jordan, is today surrounded by miles of arid desert. The evidence of its decline is preserved in an unusual way. Hyraxes (small mammals about the size of a guinea pig) live in communal mounds. They have the singular habit of cementing their homes together with urine, which dries to form an unpleasant but effective glue. It also preserves the seeds upon which the ancestral hyraxes fed. At its height, Petra was surrounded by forests of cedar and pine. These were burned. Grassland followed and this was intensively farmed. Within a few centuries, the desert had taken over. No doubt, the inhabitants of Petra in its last days fled the city, taking their genes with them.

Unfortunately no one has studied the patterns of genes in today's inhabitants of Iraq or of Jordan (some of whom may be direct descendants of the Sumerians or the people of Petra). When they do, the genetic relics of these early historical events may reveal themselves. Other early societies—and the divisions between them—have left biological traces which persist today.

Soon after the collapse of the Sumerians, the Greek *polis* or city-state appeared. Its philosophy—and its name—is at the basis of modern politics. The *Iliad* and the *Odyssey* are accounts of wars among the first *poleis*, which included Corinth, Sparta and Athens. Shortly after their foundation, Greece entered its classical age. This was a triumph not only in artistic but also in economic and political terms. By three thousand years ago Greece was the most densely populated country in Europe. Its enterprising people expanded to form Greater Greece, Magna Graeca, an empire which extended from the Caucasus to Spain. Forty towns in southern Italy were Greek. They included Syracuse, then the biggest city in the world, and Sybaris, whose wealth was a byword.

The expansion had biological effects which can still be seen. Patterns of blood groups and enzymes show that today's southern Italians and Sicilians are genetically distinct from their compatriots to the north and—because of their history—share many genes with the population of modern Greece. The genes of the first European states remain as witnesses to their past. Sardinians, too, owe some of their distinctiveness to an ancient nation-state. They are closely related to modern Lebanese, whose country occupies the territory of the Phoenicians, once the greatest traders of the Mediterranean.

The modern Greeks, unlike the Sumerians or Phoenicians, are still around as a reminder of the past. At about the time of the Greek Empire there was another buoyant economy in central Italy. This was that of the Etruscans, now a byword for obscurity. They lived in cities of up to half a million people and were skilled metal workers with, according to their Latin neighbors, a feminine and dreamy personality. Dreamy they may have been, but for a brief period their empire encompassed Rome itself. Almost no relics are left. There is the word "Tuscany" (which refers to their homeland), some enigmatic sculptures with a characteristic Etruscan smile, and a few inscriptions. There is also an eccentric object, a bronze sheep's liver covered in messages, which was used as a crib by the priest as he disemboweled the sacrificial lamb. These hints from the past were, until recently, all we knew about the Etruscan nation.

Now we know that its heritage has not been lost. Between the rivers Arno and Tiber—encompassing modern Umbria—is a region genetically distinct from its neighbors. It retains some of the genes of the Etruscans. Their biological legacy lives on in their descendants, although their language and culture have disappeared.

Religion can be as effective a barrier to the movement of genes as can nationalism. The Jews began their forced emigration from the Eastern Mediterranean in 600 BC. Soon they were scattered through Europe and the Middle East. The dispersal continued for centuries. Jews were expelled from Spain in 1492 and moved *en masse* to Turkey. Comparison of the genes of modern Jews with non-Jews from the same country (or, in the case of Turkish Jews, with Spaniards) shows that the Jewish population of each country retains some genetic distinctiveness. Although the differences are small (as are those among most populations of European origin) most Jewish peoples group together genetically. There are exceptions. In the Yemen Jews and Muslims are indistinguishable; but this probably reflects a history of conversion of the local people to Judaism. Some converts—the Falasha of Ethiopia, for example—are genetically very different, although their commitment to Judaism is undoubted. There are enough similarities between the Jews and non-Jews in most of Europe to suggest that there has been some exchange of genes since the Diaspora. The religious divide, although certainly a barrier, was not an impenetrable one.

The tribes of Israel have an infinite capacity to attract myth. Various peoples have seen themselves as one of the Lost Tribes (and hence as direct descendants of the Prophets). Even the British were not immune: there was for many years a society, the British Israelites (now extinct), whose obsession it was to prove the Old Testament origin of the royal family. Some of the declared adherents of the Lost Tribes may have a slightly better claim than others. In South Africa there is a people, the Lemba, who have black skin but look different from their neighbors. They have hooked noses, practice circumcision, will not eat pork, and pray in an unintelligible language. The Lemba have long thought of themselves as one of the Tribes of Israel. Their genes do show more similarities to modern Middle Easterners than do those of the other native peoples of South Africa. Perhaps they retain part of the genetic and cultural heritage of the Arab traders who have been visiting that part of the world for a thousand years: if not Israelite, the Lemba may at least be able to claim some Semitic ancestry.

Many other movements involved commerce rather than conquest. Often, the traders left genetical calling cards. The Silk Road passes from the ancient Chinese city of Changan to the Mediterranean. It has been a trade route for more than two thousand years and for much of this time was the main artery of cultural exchange. Silk passed from east to west; in return came cotton, pomegranates and Buddhism. Modern China has few of the genetic variants in hemoglobin, the red blood pigment, which are common elsewhere in the world (see p. 37). The blood of the peoples of today's Silk Road reveals a trail of variant hemoglobin genes which originated around the Mediterranean and spread, with the traders, along this ancient trackway. At its western end in China, about one person in two hundred carries an abnormal hemoglobin, while at the distant eastern end this drops to less than one in a thousand.

Other dispersals involved forced mass migration. The Jewish Diaspora has its parallels in the more recent past. Stalin moved thousands of people from the Crimea to Central Asia; and there are now movements of minorities across Eastern Europe following the collapse of their Communist regimes. Some of these migrations may have genetical consequences which will be seen by future anthropologists (although in the Balkans at least it seems that past turmoils have already led to so much genetical blending that ethnic boundaries—cause of conflict though they are—do not reflect genetic change).

Often, there has been a trend toward reuniting peoples fragmented by history. Israel owes its existence to the urge to reverse the damage to Jewish nationhood done by the Diaspora. In the 1920s there were exchanges of Greeks (whose remote ancestry could be traced from Magna Graeca itself) with Turks who found themselves marooned in modern Greece. Greek-speakers, many of whom may have incorporated Byzantine genes, were moved from as far east as the Caucasus.

Since the Second World War there has been a less conspicuous movement by the German government to restore its minorities marooned in Eastern Europe to their ancestral homeland.

In all these cases, part of the drive toward nationhood is the desire for unity among peoples sharing a common culture. This is often manifest in a common language, which explains the pressure on those returning to Israel to learn Hebrew and on ethnic Germans from the east to speak their ancestral tongue. As usual, Dr. Johnson put it well: "Languages are the pedigree of nations." The potency of language in defining a nation and preventing intermarriage is perfectly illustrated by the Statutes of Kilkenny in 1367. At that time, the English had subdued only that part of Ireland around Dublin which was known as the Pale. Everything beyond it was seen as barbarous. There was alarm by the authorities at the encroachment of the native peoples, which went as far as marriage with the English settlers. The Statutes declared that ". . . now, many English of the said land, forsaking the English language, manners, mode of riding, laws and usages, live and govern themselves according to the manner, fashion and language of the Irish enemies, and have also made divers marriages and alliances between themselves and the Irish enemies, whereby the said land and the liege people thereof, the English language, the allegiance due to our Lord the King, and the English laws are put into subjection and decayed, and the Irish enemies exalted and raised up contrary to reason . . . Therefore, if any Englishman or Irishman dwelling among the English, use Irish speech, he shall be attainted and his lands go to his lord." The Irish government still struggles to save the almost extinct Irish speech of the Gaeltacht; and, in a matching historical obsession north of the border, not until 1992 was the ban on using Gaelic street names in Northern Ireland lifted. For six hundred years there has been an attempt by the two nations who share a small island to retain their social isolation by using language; an attempt which, bizarrely enough, has survived the death of one of the languages involved.

Any entity—be it a language or a pool of genes—which remains isolated from its fellows will begin to evolve away from them. There are parallels between the processes of biological evolution and those which produce new languages from a common ancestor. Now it seems that there is much more than analogy between linguistic and biological change. Language barriers slow the movement of genes, and barriers between them may mark a genetic step. What is more, family trees of language look so similar to those based on the sharing of genes as to suggest a common history.

There are some five thousand different languages in the world and many more—like Etruscan—which are extinct. Like genes, languages evolve because they accumulate mutations. Some words change quickly while others are more conservative. Although there were claims in the nineteenth century that within a hundred years English and American would be mutually unintelligible, most languages retain enough of their identity for long enough to be, like genes, clues about the past and about the history of barriers to mating.

Sometimes the linguistic barriers are scarcely noticeable. England can be divided into zones defined by whether people do or do not pronounce the final letter "r" in words such as "car." I don't—I say caH; I was brought up in Wales and on Merseyside, but many of those in Cornwall, Lincolnshire or Northumbria (and plenty of Americans) pronounce it as caR. This seems trivial. However, such tiny differences can mount up until there is a barrier to the exchange of information and a new language—and often a new people—is born.

We see this throughout Europe. In Italy there are several dialects, some of which trace certain words to their Greek past. Dialects reflect history on the larger scale, too. A Portuguese farmer can no more understand a Venetian than we can; but he can talk to his Spanish neighbor, who can converse with his Catalan cousin, who in turn is linked to Italy through the *langue d'oc* in southern France. The chain of dialects reflects a history of shared descent tracing back to the Roman Empire and before.

It is sometimes possible to guess at what the ancestral languages sounded like. Father, *padre* and *père* are obviously related terms. They all descend from the same word, which sounded something like p'ter; and means that the phrase "God the father" can appear as both *deus patris* and *Jupiter* or, in Sanskrit, as *diu piter*.

A family tree of European languages shows that nearly all are related. This group, the Indo-European family, also includes Indian languages such as Bengali and extinct tongues like Sanskrit. Its existence was recognized by Sir William Jones in 1786, who saw that Greek, Latin and Sanskrit "all sprang from some common source which perhaps no longer exists." Finnish, Hungarian, Turkish and Maltese belong to other language families, but half the world's population now speaks an Indo-European tongue as its first language.

A political map of modern Europe shows lots of national barriers. A map of its languages looks generally similar. Most Frenchmen speak French, and most Germans German. Language is a clue about history and a force for national cohesion. It is also a barrier to the movement of people. It reduces the chance of marriage and the spread of genes and has done so for thousands of years. In the Bible we learn the fate of an Ephraimite prisoner taken by the Gileadites: "The men of Gilead said unto him, Art thou an Ephraimite? If he said nay, then said they unto him, Say now Shibboleth; and he said Sibolleth: for he could not frame to pronounce it right. Then they took him and slew him."

A genetic map of Europe shows that boundaries between languages can be regions of genetic change. In Wales, there are genetic differences between Welsh and English speakers. This is particularly noticeable in what used to be Pembrokeshire. This is still referred to as "Little England beyond Wales," because most of its people are English-speaking. In 1108, King Henry I moved a group of artisans there from the banks of the Tweed to set up a weaving industry. Their new anglophone homeland ended at a sharp boundary, the Landsker. Even a century ago, this was a barrier to mating, with only one marriage in five hundred taking place across the divide. Eight hundred years after they arrived, the blood groups of the descendants of the immigrants still differ from those of their Welsh-speaking neighbors.

In the same way, the population of Orkney—whose native language is a Scandinavian one—is genetically different from that of the rest of Scotland. Even dialects may mark biological barriers. In France there is a small genetic step between those who speak the *langue d'oc* (southern French) and speakers of the northern *langue d'oïl*. Genes and language tell the same story about history.

The concordance is not always absolute. The Balkans have had, and retain, a tumultuous history of movement and conquest which has obscured any relationship between linguistic and genetic units. The Hungarians, too, speak a very distinct language, although they are biologically close to their neighbors. The Magyar conquerors from the east imposed their language on their subjects, but were too few in number to make much impact on their genes. In some places there are genetic discontinuities within groups who speak the same language. The east of Iceland is genetically somewhat different from the west, although there is no language difference. This may be a relic of the history of settlement of western Iceland by Scandinavian immigrants who brought wives and servants from Ireland.

Just like their genes, the language of the Basques seems to be unrelated to any other. The Latin author Mela in the first century wrote of his bafflement at the names of peoples and rivers which meant nothing in any tongue known to him. Francis Galton himself, who often went on holiday to the Basque country, recalled "the legend that Satan himself came here for a visit. Finding after six years that he could neither learn the language nor make the Basques understand his, he left the country in despair." Satan's problem—which is shared by most of us—is illustrated by a typically impenetrable Basque proverb: "Oinak zewrbitzatzen du eskua, eta eskuak oina" ("the foot serves the hand and the hand serves the foot"). Basque may be the last remnant of the speech of Europe before agriculture. Its only apparent relative is spoken by some of the isolated peoples of what was until recently Soviet Georgia. Many Georgians believe that their own language was taken to the Basque country by Tubal, grandson of Noah, and moves are afoot to find a Basque to succeed to the restored throne of Georgia.

Safe in their mountains the Basques resisted being absorbed by the invaders, so that their ancient language, a language of hunter-gatherers, lives on. The skeleton of Cro-Magnon himself was found in a part of France which (judging from place-names) was once in the Basque country. It is not impossible that there is a linguistic, and perhaps even a genetic, link between Cro-Magnon, one of the first Europeans, and today's Basques. This last remnant of a European hunting economy is under threat. Today, Basque genes stretch for much further than the language: east to Zaragoza, now a Spanish city, and north into France. Their economy was destroyed long ago. Now their language (which is spoken by half a million people) and their culture may finally be squeezed out by modern society. Like the Etruscans—who also spoke a non-Indo-European language—soon only their genes will be left.

Of course, there are plenty of instances where genes tell us more about ancestry than does language. Genes persist far longer and can say more about the past. We see this in the Etruscans and the Basques and, over a much shorter time, in modern Americans; whose forefathers come from Europe, Africa, India and even China, but whose language is overwhelmingly English. No doubt books, films and television will in time blur the links between genes and language, but we are still in a phase of history where enough remains of the linguistic past to make it possible to guess about its evolution. Some of the guesses hint at the very beginnings of speech and perhaps at the origin of modern humans themselves.

Where did Indo-European languages come from? The first recognizable member of the group was Hittite, written in cuneiform and spoken in Turkey four thousand years ago. Modern Indo-European tongues can sound very different. "Our father, who art in heaven" is "Ein Tad, yr hwn wyt yn y nefoedd" in Welsh, "Patera mas, pou eisai stous ouranous" in Greek, "Otche nash, suscij na nebesach" in Russian and "He hamare svargbast pita" in Hindi.

Nevertheless, it is possible to find words in common for widely used objects and to use these to guess where the languages originated. There are several shared terms for domestic animals and crops. The ancient Indo-European term for sheep, *owis*, has been inferred from the Latin *ovis*, Sanskrit *avis* and English ewe. Cow was *kou*, and water *yotor*. There are similar words for corn, yoke, horse, and wheel, too.

Perhaps early Indo-Europeans were farmers, who brought their language with them as they spread. Not everyone agrees with this idea, and even where the Indo-Europeans lived is uncertain. No doubt their language started long before the first record was preserved. Some anthropologists believe that they represent a wave of invasion of Kurgan people from the Pontic steppes, north and east of the Black Sea. This area includes the lands of the Sredny Stog culture, the first horsemen, and the invasions began about 4500 BC—long after the beginning of agriculture. Another view is that the Indo-Europeans invaded much earlier and brought farming with them as they migrated from Asia Minor (including modern Turkey) three thousand years before the Kurgans. The genetic evidence does not point clearly either to Turkey or to the steppes as the source of the Indo-Europeans. Perhaps some of the very different Indo-European peoples—and languages—who make up much of modern Europe began to diverge from each other before they began to move from their homes in the east. If this is true it is going to be difficult to trace just who, if anyone, among today's nations and tongues are the ancestors of modern Europeans.

Language, archaeology and genes all bear witness to an invasion of Europe from the east. The tie between the movement of farming, genes and speech and the competing nations we see today is obvious elsewhere in the modern world. Often farming moved into an empty— or scarcely populated—land rather than into a flourishing hunting economy as in Europe. Rice growers of the Far East around the Yangtse basin took their language as well as their genes with them as they filled the Pacific. At one time this Austronesian language family was the most widespread of all, spreading from Madagascar to Hawaii and Easter Island. In Africa, farmers spread south, filling western and southern parts of Africa with Bantu speakers.

Today's most important technical advances, from the printed book to the mobile telephone, involve new ways of making contact with people. It is this which is leading to the erosion of the nation-states which have shaped history since the days of the first farmers. We can now speak to anyone in the world as soon as they can get to a telephone. New work on *global* patterns of language suggests that the first social breakthrough of all may also have involved a piece of communication technology.

The patterns of genetic change which have built up through mutation over the past hundred thousand years can be used to make a family tree of the peoples of the world. Africans form a distinct and ancient branch of the lineage. American Indians group together with their Asiatic ancestors, and Australia and New Guinea are a separate offshoot. A family tree of languages can be made in the same way. English, German and Bengali cluster together into the Indo-European family and Chinese and Japanese into a different group. A language tree based on a very few words—one, two and three; head, ear and eye; nose, mouth and tooth and so on—looks very like one made using a much more complete vocabulary. Such limited word lists are used to classify less well known tongues (such as those of Africa or the New World) with great success.

There is a new and controversial claim that all the languages of the world can be classified in this way into just seventeen distinct families, with the thousand or so native languages of the Americas falling into only three; Eskimo-Aleut in the far north, Na-Dene in southern Alaska and Canada and all the others south to Patagonia as a single group, Amerindian. The wide distribution of this single family contrasts with the pattern in Papua New Guinea, where in a much smaller space there are nearly eight hundred languages, many of which are almost unrelated to each other. It is more than a coincidence that the genetics of the speakers of the languages of the Americas and of Papua New Guinea shows the same pattern: the Americans are genetically rather uniform, while the Papuans vary greatly from valley to valley. The highest concentration of language diversity is in the Caucasus, between the Black and Caspian Seas. In an area only twice that of Britain there are forty very different languages, some spoken in only a single village. There are even hints of a link with the American Na-Dene group and with Tibetan. Unfortunately we know nothing of the genetics of the Caucasus.

We can make a tree showing the relationships of all the world's languages and even guess—wildly—at some of the original words at the base of them all. Some Russian linguists have attempted to reconstruct Nostratic, the twelve-thousand-year-old language thought to be the ancestor of Indo-European and its relatives. These include the Elemo-Dravidian family spoken in parts of India, the Altaic tongues which include Turkic and Mongolian and an Afro-Asiatic group spoken in the northern half of Africa. They have reconstructed over a thousand "root" words. *Tik*, for digit, finger or toe, is one of these, *kujna* for dog another. There are no widely shared words dealing with farming, suggesting that this proto-language may indeed derive from a time before agriculture.

Amazingly enough, when the world language tree is put next to the genetic tree, they look rather similar. Both come to the same root in Africa and both show the same split between Australasia and other Asian peoples. Perhaps this shows that language itself dates back to the very beginning of humankind. Speech marks a huge jump in the speed of information transfer. If we were to spell out this sentence, letter by letter, it would take ten times longer to transmit the information than by speaking it. The plight of the deaf and dumb in modern society shows how much life depends on an ability to speak. It is hard to imagine any society which could work without it. The bones of early modern humans show that there were changes in skull shape and in the position of the larynx which might have marked the earliest ability to frame an articulate sound. This, and the concordance between the genes and the languages of today, suggests that it may have been speech which made humans human in the first place.

Shelley felt as much: in *Prometheus Unbound* he has his hero "give men speech, and speech created thought." Not everyone agrees with him. Some anthropologists suggest that even Neanderthals had quite a sophisticated language which (presumably) disappeared utterly when they themselves became extinct. There is a hint of a much earlier dawn of language. Apes in groups spend much of their time grooming, to show their fellows that they belong. If early humans reassured their companions as apes do, they might—because of the size of each hunting band—have had to spend half their time grooming. Speech, even primitive speech, is a much better way of calming one's fellows than is touch. The very first words—long before the ancestors of today's languages—may have been words of comfort.

Nobody will ever be able to reconstruct and to speak Neanderthalish, if it existed. There have been many claims as to the language of Eden. The sixteenth-century German philosopher Becanus was convinced that it was Old German, and that the Old Testament had been translated from this into Hebrew (although the Holy Roman Emperor Charles V said that he spoke French to men, Italian to women, Spanish to God and German to horses). Soon, there may be a chance to find out the truth. The fossils and the genes have already given us clues about where and when Adam met Eve; before long, we may even be able to guess at what they said to their errant offspring.

10 | Darwin's Strategist

AMERICAN BIRD-WATCHERS know that the common sparrow—the bird that hops around in English gardens—has a bigger body and shorter legs in the north than in the south of the United States. The same is true for sparrows in northern and southern Europe. Creationists (and there are more than a hundred million of them in the U.S.A.) see in this the deity arranging things so that each species fits into the economy of nature; cold places, wherever they are, meriting a subtle change in God's plan.

If there is a plan, it seems to work in the same way for humans. People from the far north have shorter arms and legs and more compact bodies than those from the tropics. That is why Olympic long-distance running records tumbled after East Africans with their long legs began to take part. To philosophers before Darwin the ability of Africans to cope with heat and Eskimos with cold was excellent evidence for divine action. The Creator had seen to it that each people suited their homeland, showing just what a wonderful designer he was. As the nineteenth-century cleric William Paley argued, if one found a watch, beautifully designed, then one must accept the existence of a watch-maker. The perfection of humanity proved in the same way that there was a God. This idea seemed such a powerful one that it was carried to absurd lengths. Voltaire, in *Candide,* parodied it with Dr. Pangloss and his delight at seeing the perfection with which noses had been de-signed to carry spectacles. Freud, a keen Darwinist, commented that one might just as well argue that the fact that cats have two holes in their skin precisely where their eyes are could be explained in the same way.

Now we know that there is a big problem with the argument from design, as it is called, for the sparrows at least. In fact, English sparrows have not been in the Americas since the time of creation. They got there little more than a hundred years ago. A few were brought from England and released in Brooklyn in the 1850s. They spread to fill the continent, taking about a century, a hundred sparrow generations, to do so. How did they come to resemble so closely the sparrows of their native land?

The answer lies in natural selection: inherited differences in survival and reproduction. Studies of marked sparrows in Kansas show that large individuals with short legs survive better in freezing weather. They hence have a chance to breed and to pass on their genes when spring comes. The birds released a century ago brought from their native land genes for large or small size and stocky or graceful legs. In the north, the big squat birds did better, but in those that spread to the torrid south the opposite was true. In a few generations, American sparrows evolved just the same geographic patterns as those in Europe. Natural selection had done its work.

Natural selection was Darwin's Big Idea. It was a mechanism which drove evolution and led to change without the need for a designer to supervise every step. *The Origin of Species* starts with a long section about farming. It discusses the way in which domestic animals emerged from their wild ancestors because of preferences, often inadvertent, for one type of offspring over another by those who breed them. Selective breeding, the choice of the best animals to produce the next generation, led to the divergence of new forms from the wild stock.

Darwin suggested that if farmers could do so much in a short time, then nature could do a great deal more. "If man can by patience select variations most useful to himself, should nature fail in selecting variations most useful, under changing conditions of life, to her living products? What limit can be put to this power, acting during long ages and rigidly scrutinising the whole constitution, structure and habits of each creature—favouring the good and rejecting the bad? I can see no limit to this power, in slowly and beautifully adapting each form to the complex relations of life." Darwin came to his idea by reading the work of Malthus, who pointed out that populations will always, unless checked, grow to exceed resources. Oscar Wilde put it more succinctly: "Nothing succeeds like excess."

The engine—if not the engineer—of evolutionary change is natural selection, the preservation of favored types in the struggle for life. Change is inevitable in any system, be it genes or language, in which there are errors of transmission from one generation to the next. Although change of this kind is certainly evolution, it is evolution at random. It cannot lead to progress from simple to complicated of the kind which Darwin was interested in and which gave rise to humans from their modest predecessors. Natural selection takes advantage of the fact that, each generation, inheritance makes mistakes. Because some are better at coping with what life throws at them they copy themselves more successfully. Darwin's mechanism sorts out the best from what mutation supplies. It gives a direction to evolution and allows a living system to escape from the inevitability of extinction. This is as true for humans as for any other creature.

Selection is a simple idea. As Thomas Henry Huxley, Darwin's contemporary, said when he first heard of it, "How very foolish not to have thought of that!" Exactly the same notion is now used by engineers. When making a turbine blade or a spray nozzle, they do not spend hours trying to work out the best possible design from first principles. Instead, they make a guess at what might work, try it out, and then make some new versions with slight changes. These are then tested. The best are chosen, mutated again, and the process repeated until an efficient blade or nozzle emerges. Often this has a complicated and unexpected shape which could not have been designed even by an expert engineer.

The computer programmers who use the same approach dare to refer to their creations as "artificial life." By programming their toy not with the precise details of what is needed, but with a guess of what might work, allowing this to make rough copies of itself, and choosing the most successful, they make rapid progress. In just a few generations there may appear computer birds which flock just like starlings, mathematical ants which follow trails and programmed flowers as beautiful and unexpected as anything in nature. There is even a school of computer art, in which the agent of selection is the artist's perception of beauty. Each generation a new set of shapes appears. By choosing the most pleasing and going through the cycle again and again, bizarre and eccentric images emerge. Literature is not immune to the advance of what might be called artificial natural selection. Analysis of simple works such as medieval folktales or children's books show that just a few unpretentious themes underlie them. Feeding them into a computer, changing each motif slightly, and choosing the best means that brand new and artistically satisfying medieval tales or children's stories emerge quite quickly.

Humans are certainly not safe from the action of the Darwinian machinery. For most of history, most of us died before we were old enough to pass on our genes. Even among the survivors, some had more children and some fewer. If any of these differences in survival or reproduction are influenced by genes, then Darwin's mechanism is at work. The constitution of the next generation will differ from that of its parents. There has been selection and in time there will be advance.

The power of the evolutionary machine rests on its ability to choose the best available at a particular time, even if it is only a little better than what went before. Lewis Carroll illustrated how it works. Imagine that we have a three-letter word—"pig," for example—and we want to change it into another—"sty." We can change any letter into any other. If we make random changes and just hope for the best, taking any meaningless set of letters each time, it takes thousands of moves to get the pig into the sty. Natural selection imposes a rule: all the words in between must make sense. It picks up the combinations which look good and builds on them. It can get there in just six steps—pig, wig, wag, way, say, sty.

Wherever we look, we find selection. The living world is full of it, because the living world has been formed by it. Its actions can sometimes be subtle. As well as inherited differences in survival there may be genetic differences in the ability to find a mate or to produce offspring.

The theory of evolution was a sensational one in 1859, the year *The Origin of Species* was published, because it seemed to remove the need for a direct link between God and man. There is a story of one Victorian lady (the wife of the Bishop of Worcester) saying to another about Darwin's work: "My dear, let us hope that it is not true—but, if it is, let us pray that it does not become generally known!" After the Church establishment had recovered from the shock, several religious thinkers came up with the idea that evolution was in fact a way of working out God's plan. Even if humans were not perfect, they were perfectible, and natural selection was the means which the deity had chosen of doing it. However, its action, far from perfecting the imperfect, often seems incompetent or even cruel. There is not much comfort for Panglossians here.

Selection can do astonishing things. But some things are beyond it. Natural selection cannot plan ahead; it acts without foresight, taking no thought for the morrow. Not only does it do just what is needed and no more, but sometimes it does it in what seems like a slapdash and shortsighted way. It is, to use Richard Dawkins' memorable phrase, a blind watchmaker, achieving a remarkable end through a simple and inefficient means.

There is a danger of seeing the whole of biology as evidence for natural selection. Just as Paley saw the complexity of life as an argument for God, there is a neo-Paleyism which plagues evolutionary theory. It argues that all animal structure is well adapted so that it must always reflect the action of selection. This argument may be circular, but is difficult to disprove. It has led to many disagreements—fascinating to their proponents, tedious to those outside—among biologists. Some feel that the Darwinian machinery drives the whole of evolution from the order of the bases in the DNA to the shape of the nose. Others see selection as just an occasional event which directs some genes while most change at random. The issue remains unresolved.

Another beauty—and an important weakness—of the theory of evolution by natural selection is that with a little imagination it is possible to come up with an explanation of anything. Evolutionary biologists like to spend their time making up stories about how selection has molded the most unlikely characteristics. Sometimes they even turn out to be right. Anthropologists have particularly vivid imaginations. There are some startling guesses about how it may have formed human attributes. Many are fantasy, but because they invoke events which took place in the distant past they are almost impossible to refute.

It may also be the case that there are vast areas where differential survival and reproduction is working without anybody realizing. Only one egg in thousands and one sperm in millions actually produce an offspring. Do the others die at random—or do they fail for genetic reasons? Nobody knows: but if only the best among them survive, the Darwinian machine is a far more pervasive force than was ever thought.

Whatever its general importance, selection is just a mechanism and not a force for good. Cancer patients are sometimes given a drug which attacks dividing cells. The treatment often fails. Natural selection is at work. A few cells have undergone a mutation which changes the properties of a certain gene to enable it to break down the drug. These reproduce more rapidly than the others and soon take over, sometimes so effectively that the patient dies. There is not much evidence of a benign designer here.

Although we often have no idea why a particular character has evolved, humans show as well as any other creature the strengths and weaknesses of selection. They changed quickly, adapting as they filled the world over the past 150,000 years. This represents around six thousand human generations. The same number of generations of mice goes back only to those which infested the newly built Acropolis and of fruit flies to the insects swarming over the apples of William the Conqueror. As far as anyone knows, today's mice and fruit flies have scarcely changed over that time, emphasizing just how fast human evolution has been.

There have been three ages of human history, each of which has molded, and been molded by, the action of natural selection. A lengthy Age of Disaster was followed by a shorter Age of Disease and—very recently—by an Age of Decay. For most of the past nearly all those born died disastrously, through cold, hunger or violence. Many individual tragedies acted as agents of evolutionary progress. This chapter is about how humanity evolved to cope with the changes in climate and diet as we moved from our African home. The second epoch, the Age of Disease (which began only a few thousand years ago) although largely over in the West is still very much around elsewhere in the world. Disease is such a potent agent of natural selection that it deserves a chapter, the next one, to itself. The Age of Decay (during which most people die of old age) is now upon us. Because most people who die nowadays have passed on their genes it is hard to know what selection is going to be able to do about it.

Our ancestors, our relatives and ourselves are tropical animals. In spite of Noël Coward, humans are one of the few large mammals that can cope with the African midday sun. Most people, given a choice of habitat, prefer a warm place (even if it is only Palm Beach for a couple of weeks a year) and many adaptations have evolved to combat heat, rather than cold. Humans are the least hairy of all primates and sweat the most. On a sunny day, the temperature at, or a few inches above, the ground surface can be as much as twenty degrees higher than that only a couple of feet away from the earth, because the ground absorbs and reflects heat from the sun. An upright posture, which did so much to differentiate humans from apes, may have evolved in response to high temperature. One of the best ways of cutting down heat stress on sunny days is to stand up, out of the layer of hot air near the surface. Perhaps, as our distant ancestors moved into the savannahs from the forests, they stood up to cool down; opening, literally and figuratively, new horizons for their descendants.

Humans today live in every environment from rainforest to tundra and from sea level to fifteen thousand feet and more above it. Culture —fire, clothes and houses—has helped, but there have been genetic responses to climate as well.

We left Africa more than a hundred thousand years ago and reached New Zealand, the farthest point of our spread, only a thousand years before the present. For much of the time, the weather was even worse than it is today. Ancient climates can be inferred from shifts in the chemical composition of water. In the Arctic, this falls as snow and is preserved as ice. A core has been drilled through ten thousand feet of Greenland ice to reach the rock below, where the first snows fell two hundred thousand years ago.

The record of the icecap shows that there have been many ice ages during the evolution of *Homo sapiens*. The last peaked about eighteen thousand years ago. It had a drastic effect on the large mammals of the world, humans included. The giant sloths and native horses went from the Americas, the woolly mammoth from Asia and giant lemurs from Madagascar. Large areas of northern Europe which had once been populated were abandoned. As the climate dried up because water was trapped in the ice, parts of Africa became desert and were lost to habitation as well. The level of the sea fell as water was locked into the ice. The Bering Strait and Bass Strait dried out. Broad coastal lowlands emerged in many parts of the world. The air filled with dust—still preserved in the Greenland ice—from the freezing deserts. Even if they were cold most of the time, our ancestors had beautiful sunsets.

In the Russian Plain there were settlements within a hundred and fifty miles of the icecap. The early Frenchmen who painted the cave at Lascaux could not relax by basking in the sunshine in a pavement café. The arctic ice was only three hundred miles away and they had to stay warm to stay alive. Perhaps the need to keep under cover fostered artistic endeavor. The explosions of style in painting and tool-making all happened on the chilly edges of humanity's range rather than in the tropics. Humans survived the harsh new climate and at the peak of the last glaciation were the most widely distributed mammal in the world —a status they have retained ever since.

Not all was gloom during the global spread. There were brief periods, up to a couple of thousand years long, when the average temperature rose by as much as seven degrees centigrade in just a few decades. This is a dramatic change, equivalent to the Scottish climate shifting to that of southern Spain within a single lifetime. Perhaps these sudden strange warmings impelled the colonists on their way.

Just as in the sparrows, differential survival and reproduction favored those best adapted to climate. The Neanderthals, our extinct cousins, who had lived in a chilly Europe long before the modern upstarts arrived, were short, squat and heavy and hence well adapted to the cold. Most people would change seats if Cro-Magnon, an early European, sat next to them on the subway, but would change trains if a Neanderthal did the same thing.

Modern humans show geographical trends in body build which also reflect the action of climatic selection. Eskimos are about a third heavier for a given height than the world average, while men from some East African groups are much slimmer than other peoples, at about three quarters the weight expected for their height. Much of this difference arises from changes in body proportions. Most peoples from the tropics are tall, thin and have long arms and legs. Those from the north tend to be more heavily built. For unknown reasons, the trends are stronger in men than in women. The same is true for the trends in body shape in sparrows, perhaps because larger males are more aggressive when fighting for food in wintry conditions. Although little is known about the inheritance of characters such as size and shape (and although there are certainly some environmental effects) the differences are at least partially genetic.

The short fat peoples of the north are better at conserving heat in the body core. Those with more graceful figures from hotter climes cool down more effectively as they lose heat through their long arms and legs. This pattern exists in other animals, from birds to foxes, and the trends in body shape have been graced with a name of their own, Bergman's Rule. Most of the body's excess heat is lost from the skin and the amount of skin surface per unit of volume is greater in thin and spindly individuals.

There are more subtle differences in the control of heat loss. Some populations are better at regulating the amount of heat that gets to the arms and legs through blood vessels in the skin. If a European or an African puts a finger into icy water, its temperature quickly drops to a level low enough to damage the flesh. When an Eskimo does the same thing his finger stays reasonably warm. Again, it is not clear how much of this effect is genetic, but among North Atlantic fishermen those of European origin are worse at keeping their hands warm than are the Eskimos. Australian aborigines have another way of dealing with a climate which is hot during the day but chilly at night. They close down blood vessels near the surface on cold nights, so that their skin temperature falls to well below that of a European facing the same conditions, saving heat in the body core. Aborigines are also better able to tolerate cold without shivering. All this means that they can sleep in the open without too many problems. Even the rate at which the body uses energy is lower in those who evolved in the tropics.

Other patterns might also be due to climate. The woolly hair of Africans is said to act as an evaporating surface for sweat to cool the head down. The long fine noses of peoples from the Middle East may help to moisten the desert air before it reaches the lungs and the narrow eyes of Chinese to protect against the icy winds of the Asian plains. All this is guesswork.

There is one apparent anomaly in the global trends in response to climate. This is the pattern of skin color. In the Old World at least, most tropical peoples have darker skins than do those from cooler climes. As anyone who has sat on an iron park bench on a sunny day knows, black objects heat up more in the sun than do white, so that black skin, far from protecting against the sun's heat, actually soaks it up.

There are several theories as to why humans evolved light skins as they migrated to the dismal climates of the north. None is completely satisfactory. Skin cancer brought on by the ultraviolet in sunlight is a possibility. Malignant melanoma is a dangerous cancer which is particularly common in countries such as Australia where people with light skins expose themselves to high doses of ultraviolet by sunbathing. The new habit of lying in the sun for fun has led to a rapid increase in its incidence. Until recently, when the dangers of sunburn became widely known, the rate of skin cancer among whites was doubling every ten years. Those with light skins are in most danger, people with fair complexions and red hair most of all. Black people rarely get the disease. There is one exception: in parts of Nigeria there are many albinos (known, in some places, as "DOs"—District Officers—with cynical but inaccurate reference to their probable parenthood). Nearly all of them get some form of skin cancer during their lives.

Even so, cancer has probably not produced the global trend in color.

First, it is rare even in whites, with only about one case per ten thousand people per year. More important, skin cancer is mainly a disease of the old. This means that those who die from it have already passed on their genes, including those for the color of their skin.

Global trends in color may be an indirect response to climate. Without vitamin D, children get rickets. Their bones are soft and deform easily. Ancient graves show that this has been a problem for thousands of years. Rickets was still common in the slums of Victorian England. There is vitamin D in milk and most European children are now protected from the disease by a healthy diet. Vitamin D can also be made in the skin by the action of ultraviolet light. People with white skin do this more efficiently than do those with black. Under a UV light, whites synthesize a useful dose in only half an hour, while blacks take three hours to do so. Even a few hours in sunshine allows a light-skinned baby to make enough vitamin D to avoid rickets and it is no accident that African babies are lighter in color than are adults. For the same reason, rickets is more common among immigrant children in Europe and blacks in the U.S.A. than among their light-skinned counterparts. Natural selection acted to favor those with genes for light skins as humanity began its long walk from the sunny tropics to the gloom of northern Europe.

Why black skin is common in the tropics is less clear. Although vitamin D is harmful in large amounts even fair-skinned people cannot make enough in sunshine to do any damage, so that dark skin pigment has not evolved for this reason. Perhaps a black skin stops sunlight from destroying other vitamins in the blood as they circulate through the outermost layers of the body. There is some evidence for this. Patients with skin diseases are sometimes treated with intense doses of ultraviolet. Those with light skins show a sudden drop in certain vitamins known to be essential for pregnancy and normal growth. The black skins of tropical peoples may help in coping with vitamin destruction, an effect which may be particularly important as there is often a shortage of vitamins in the food. Another possibility is that the dark pigment prevents ultraviolet from destroying antibodies in the blood as it circulates through the skin. Yet another is that black skin allows the lightly clothed peoples of the tropics to warm up quickly in the early morning as the sun rises, even if they have to shelter during the heat of the day—when dark skin may act as camouflage in shady places. As usual, it is easy to make up stories about how selection may have favored certain genes, but none can be taken very seriously without more experiments to see which might be true.

In creatures such as snails and fruit flies heat has many genetic effects. It acts on inherited differences in enzyme structure, increases the mutation rate, and even causes "selfish DNA" to start hopping around the genome. Humans can, of course, regulate their internal temperature quite well so that differences in climate must have a less direct effect. Nevertheless, there are north–south trends in blood groups and even in the alternative forms of certain enzymes. Whether these are due to selection by climate is not known.

Humans, like most creatures, live on a thermal tightrope. If our temperature goes up by only a few degrees, we die. Molecular biology has illuminated the imminence of thermal disaster. One group of genes produces "heat shock proteins." In snails and fruit flies these are switched on when things get too warm. Sometimes, most of the cell's machinery is devoted to the job. Humans have them too. During a fever, cells start making heat shock proteins. They cluster around delicate enzymes which might be damaged by high temperature. Even a rise in body temperature of a couple of degrees sets the protective machinery into action. Perhaps there are differences in the sensitivity of the heat shock system in peoples from tropical and temperate climes. As yet, nobody knows.

Heat shock proteins are an emergency measure. Lower animals were once described rather dismissively as "cold-blooded." They lack the bodily machinery which keeps mammals (including ourselves) warm. However, many keep their temperature more or less stable just by behaving in the right way. One species of lizard thrives from the deserts of California to the icecaps of the Andes. It keeps its temperature almost the same across this vast range of climates simply by moving in and out of the sun. I once invented a paint which fades at a measurable rate when exposed to daylight. Putting spots of this onto snail shells shows how long each animal has spent in the sun over a month or so. Snails from hot and cold places behave differently and within a population dark- and light-colored snails (which differ in the extent to which they soak up solar energy) also differ in exposure to sunshine. Perhaps the same method could be used to study dark- and light-skinned people, too.

Painstaking observation of animals in sunshine shows what a crucial role behavior plays in controlling temperature. Desert lizards cannot stray more than a couple of yards from shade or they would die of heatstroke before they got back, but they are obliged to venture into the sun every few minutes to feed. Some spiders spend half their energy budget shuttling back and forth between sun and shade. A spider living in a place with the right balance of shady and sunny patches can produce far more eggs than one whose home has plenty of food but not enough sunshine. It is easy to forget the importance of behavior in our own thermal lives. A quick estimate of what keeping at the right temperature costs the average Briton (or, even more so, the average inhabitant of Chicago)—a bill which includes houses, clothes, central heating, air-conditioning, food and, of course, holidays in Marbella or Florida—shows that the spiders are relatively modest in the proportion of the budget spent on keeping comfortable. Warm-blooded though human beings may be, evolution has forced us into some cold-blooded decisions about how to stay alive as we moved out of the tropical climates to which our ancestors were adapted.

Humans, like most mammals, are adapted to lowlands. They cannot survive for long at over fifteen thousand feet above sea level as the amount of oxygen in the air is only half that available lower down. In the Andes there are settlements at this height. The children of Andean Indians are better able to cope with such conditions than are those of immigrants from the lowlands. Even native highlanders brought up at sea level are better at extracting oxygen from the mountain air so that perhaps there has been an evolved response to oxygen starvation.

Diet, too, has been an agent of evolutionary change. In the world as a whole only a minority of adults (including much of the population of western Europe) can digest cow's milk. This is not surprising. Most creatures (including all humans before agriculture) never have the chance to drink milk of any kind after they have been weaned. Digesting milk depends on an enzyme which allows lactose, one of the milk sugars, to be digested. If this stays active until adulthood, cow's milk is a useful food. If it does not, then milk loses much of its value and an adult who drinks a lot of it suffers from wind and indigestion. The milk-digesting gene is rare in much of Africa and in the Far East (which means that the dried milk once sent as food aid for adults in these places was largely wasted). It is much more common in western Europe and in some African peoples such as the Fulani of northern Nigeria who herd cattle. Which is the evolutionary chicken and which the egg here is not certain. Perhaps the gene was first favored in desert peoples as it allowed them to drink camel's milk to get water. In Europe it may be favored as those who have it can extract calcium from cow's milk and avoid rickets. Again, it is fatally easy for imagination to take precedence over experiment.

The best-understood force of selection in humans, which works on molecular variation as much as on more obvious characters, arises from the existence of inherited differences in resistance to disease. Disease seems to be an unavoidable part of life: even creatures preserved from the dawn of existence show signs of infection. Some cell constituents (such as mitochondria) may be the remnants of disease organisms with which higher organisms have learned to coexist. The games of computer "life" based on an analogy of natural selection already have their diseases in the form of computer viruses. Disease has a history and a geography: people have faced different illnesses at different times and in different places. Infection is a relentless enemy as it involves living creatures—the agents of disease—who themselves must evolve in response to the body's defenses, or die out. There is an evolutionary arms race between ourselves and our diseases. To see what natural selection can and cannot do, and the way in which it may levy a high price on its subjects this race must be looked at in more detail. That is the task of the next chapter.

11 | The Deadly Fevers

IN A FIFTEENTH-CENTURY CHRONICLE by the first Portuguese explorers of West Africa is a bitter complaint: "It seems for our sins, or for some inscrutable judgement of God, in all that we navigate along He has placed a striking angel with a flaming sword of deadly fevers." Three hundred years later, half the Englishmen who went to that part of the world died within a year. When Europeans and their African slaves first went to South America, it was the natives who suffered. The population of Mexico dropped from twenty-five million to one million between 1500 and 1600. Some tribes disappeared altogether. The number of Quimbaya in Colombia paying tribute to the Spaniards was fifteen thousand in 1539, but sixty-nine in 1628. Everywhere, the great killer was infection: malaria, smallpox and typhus. In both New and Old Worlds those who had lived with a disease for many generations survived better. There seemed to be inborn differences in resistance between people from different places. At the time the differences seemed to be miraculous. Now, we know that the evolution of mechanisms for resisting disease is the best example of natural selection in action. Although the Age of Disease may be (at least temporarily) over for the western world its genetic consequences will persist for thousands of years to come.

When facing disease, selection is dealing with an unremitting enemy. Western society has won a respite in the battle but throughout recent evolutionary history pestilence has been the greatest killer and the greatest agent of selection. In the fourteenth century—thirty human generations ago—the population of England was halved by the Black Death. Today, there are predictions that the spread of AIDS in some African countries means that their populations too will soon begin to drop. Death from cold or starvation may be brutal, but at least the enemy is predictable. Bacteria and viruses are alive, and need a constant supply of new hosts. They can evolve as well, which means that there is a continual evolutionary race between our ability to adapt ourselves to viruses and bacteria and their ability to infect us. It is an implacable and apparently endless relay race. As soon as one opponent has been defeated, another comes along.

Patterns of infection depend on the number of people available to be infected, so that the importance of disease has changed throughout history. The longer an illness has a hold on someone, and the more efficiently it is transmitted, the smaller the population needed to allow it to persist. Immunity also plays a part. This means that some sicknesses must have started long before others. Signs of tuberculosis, which can last for decades, can be seen in the bones of those who died tens of thousands of years ago. In contrast, measles must be new. It does not last long and is not particularly infectious. Those who have been infected become immune and cannot be infected again. The history of measles is close to those of the many new afflictions which have attacked the human species since it began.

In a population which has never been exposed to it and in which there is no immunity, measles can have terrible effects. When it came to Fiji in 1875 (the result of a visit by the King of Fiji to Sydney) it killed nearly a third of the hundred and fifty thousand Fijians. However, it soon disappeared from the island as it needs a community of at least half a million people to keep going. Measles may sometimes arrive in a place (like Fiji) with fewer people than this, but cannot persist. In Iceland before the Second World War there were gaps of up to seven years between epidemics. Only after 1945, when constant movement in and out meant that the Icelanders became part of the European population as a whole, did measles become a continuous problem. Humans have lived in groups of half a million or more for only two or three thousand years, so that measles must be a fairly new disease. Its initial impact was much more damaging than its effect on populations which had lived with it for many generations.

The constant change in the pattern of infection means that evolution can never rest: far from perfecting us, it is constantly faced with new problems. Ten thousand years ago, humans lived in small bands. Contagious disease may scarcely have existed. No doubt there were plenty of lice and tapeworms around as their long lives and tendency to re-infect their own hosts mean that they do not need many people to keep going. In general the ancient world was a healthy one. People starved, froze or were eaten by tigers instead. Even when infection struck, it was a local problem. There are remnants of this pattern among the few hunter-gatherers who remain. In the 1950s, different tribal groups of Yanomamo Indians differed greatly in the antibodies which they pos-sessed. In some villages, everyone had an antibody to (and had hence been infected with) chicken pox. In others chicken pox had never arrived, but the whole surviving population had once had influenza. Each small tribe had a different history of infection. The pattern of disease was a balance between a chance arrival of a new pathogen and a local epidemic which ended as soon as everyone was immune or dead. The same pattern of life—and of death—exists in chimpanzees today. Now, the small islands of Yanomamo Indians have joined the rest of humanity, with its diseases, and are suffering as a result.

With farming, the human population shot up and began to coalesce into one continental mass. A whole new set of disorders appeared. Irrigation helped, and water-borne parasites such as schistosomes, which are carried by snails, began to appear. Their eggs have been found in mummies from 1200 BC. Schistosomiasis (which is also known as bilharzia) is still common in Egypt. Many of the biblical plagues were probably new diseases taking hold as the population of Egypt grew large enough to sustain them.

Some of the contagions came from animals. The closest relative of measles is the cattle disease rinderpest. Measles itself may have evolved from this. A close relation of smallpox is found in cows and of sleeping sickness in wild game. Only a small genetic change in the parasites was required to allow them to infect a new host, *Homo sapiens*. The same process is at work today in the waves of "Asian flu" which occasionally sweep across the world. Each strain originates in ducks on Chinese farms. It shifts to the local pigs and then to the farmers. Every few years one strain attacks millions of people worldwide and then disappears until a new mutant form emerges.

Some maladies came and went mysteriously and have never been identified. In the towns of medieval Europe there were outbreaks of a dancing mania in which thousands took part. Some may have been due to mass hysteria but as there were swellings and pain there must also have been an organic cause. In Italy the affliction was called tarantism and ascribed (wrongly) to the bite of spiders. St. Vitus' Dance, with its visions of God, was probably the same thing. The epidemics began in eleventh-century Germany and had disappeared by the seventeenth century. England had its own mysterious and transient illness: the English Sweat, which came and went several times between 1480 and 1550. The effects were terrifying. It was brought into the city of London by soldiers escaping the Battle of Bosworth and a month later was at its peak, killing nearly all it infected within a day. The University of Oxford was closed for six weeks. There were several epidemics over the next fifty years, some spreading to mainland Europe, where the mortality was so high that eight corpses were put into each grave. The last epidemic started in Shrewsbury in 1551, killing thousands. Since then the disease has gone. What it was, nobody knows.

Hippocrates, in the fourth century before Christ, was the first to describe symptoms well enough to allow diseases to be diagnosed with any confidence. Ancient Greece had diphtheria, tuberculosis and influenza, but none of his records suggest that smallpox, bubonic plague or measles existed there. Movement between the civilizations of the world led to a new set of pestilences. Smallpox was present in India a thousand years before Christ, but its short incubation period meant that it killed its carriers quickly and did not travel well by land. It reached Europe by sea, causing the first epidemic in Rome in 165 BC. Anglo-Saxon records mention nearly fifty smallpox epidemics between 526 and 1087.

Things got even worse in the first large towns. Big cities are recent things. Before 1800 only one European in fifty lived in a city of more than a hundred thousand people. There has been movement from the countryside for more than a thousand years, but epidemics meant that no city was self-sustaining in numbers until the nineteenth century. London at the time of Pepys had a population of a hundred thousand, but it needed five thousand immigrants a year to maintain its population in the face of pestilence.

Plague had killed millions in England in the centuries before Pepys' day, but its last—and worst—epidemic was during his own lifetime. In December 1664, two Frenchmen died in Drury Lane. In the following June Pepys wrote in his diary: "This day, much against my will, I did in Drury Lane see two or three houses marked with a red cross upon the door, and the 'Lord have Mercy upon Us' writ there: which was a sad sight to me, being the first of the kind that, to my remembrance, I ever saw. It put me into an ill conception of myself and my smell, so that I was forced to buy some roll-tobacco to smell and to chew, which took away my apprehension." By the summer of that year, two thirds of the population of London had fled and the disease was raging throughout England. The cycle of plague epidemics which tormented London and reached its peak in the Plague Year of 1665 ended with the replacement of thatched roofs (and their resident rats) by slates after the Great Fire in 1666. The last European plague was a century later, in the Balkans. Although the disease has often been introduced since then, it has never spread.

The control of pathogens is a recent thing. In England until only three hundred years ago there were cycles of mortality. Life expectancy fell from forty-two years in the late sixteenth century to thirty years in the seventeenth, returning to the earlier level only in Victorian times. The greatest mortality was in low-lying villages. "Fevers" were usually blamed. All this meant that city conscripts impressed into the army did better than the healthier youth from the countryside. Although the urban soldiery was pinched and weakly they had been exposed to infection so often that they were immune to the diseases which quickly slaughtered their country cousins when forced into crowded barracks.

New contagions continue to appear. As well as AIDS, Africa had another mysterious epidemic in the 1970s, when outbreaks of the deadly—and previously unknown—Ebola fever killed half those infected. Even a trivial change can spark off new illnesses. In the past decade, Lyme Disease (named after the village of Lyme, in Connecticut, where it first appeared) has become the most widespread pest-borne disease in the United States, with more than ten thousand cases a year. It causes arthritis and a variety of painful nervous symptoms. The malady is due to a microorganism which spends part of its life in a tick found on white-tailed deer. There were a few cases of Lyme disease a century ago but it became common only recently when people moved to the suburbs and were exposed to the deer which flourish there. Nineteenth-century sanitation meant that cities became far safer places. But it had a cost. Before sewers, nearly every infant was exposed to a constant small dose of polio virus. Their immune system works very efficiently and most became resistant. Once drinking water had been cleaned up only those few children unlucky enough to come into contact with a sudden dose of the virus got the disease.

For most of the world disease is still a scourge. Ten million lives a year are lost to measles, and five million to diarrhea, although these could, given the political will, be controlled by vaccines and by clean drinking water. Bilharzia, another infection which should be simple to contain, attacks two hundred million people. Faced with this onslaught by a series of changing enemies, natural selection can never afford to relax. As more is discovered about human genetics, the importance of disease, extant or extinct, looms larger. It may be that much of the mass of human variation is a remnant of past battles against infection and that many of the genetic trends across the globe result from natural selection by disease today or in earlier times.

Disease itself also evolves. If it did not, its agents would soon themselves become extinct. It was once assumed that such evolution must inevitably lead to a truce with those infected: it seemed obvious that the best strategy for a pathogen would be to keep its host—its homeland—alive. Sometimes, no doubt, this is true. However, the genes which drive a disease evolve in their own interest alone. If the most efficient way of increasing their number is to kill the patient, then evolution will provide the means to do so.

AIDS may once have been relatively benign. In Senegal—where levels of promiscuity are low—the virus does not do much harm to those who carry it. They stay alive long enough to have a second sexual contact (perhaps after many years) and in so doing to give the virus a new home. Only with sexual excess—as in East African truck drivers or gay New Yorkers twenty years ago—did the virus become as virulent as it now is. It could afford to multiply within a host in the expectation that, before he died, he would pass on the infection. What seem like social improvements may do the same thing. Before flush toilets (which, in their early days, fed directly into rivers) cholera was less dangerous. It had to keep its prey healthy for long enough for them to move to another village and to pollute its wells with the cholera bacillus. As soon as one patient could infect hundreds more as his waste poured into a river cholera turned vicious: the victim needed to survive only to reach a toilet. If he died from fluid loss while pumping out millions of bacteria, that mattered not at all.

I will concentrate on humankind's genetic response to just one infection: the evolutionary struggle against malaria. Three hundred million people are infected and the disease kills nearly two million a year —half of them African children. Almost half the world's population lives in malarious regions. One estimate is that the death rate will double within the next thirty years. Increased travel means that the disease can spread quickly. There are more than two thousand imported cases in Britain each year and, very occasionally, malaria is transmitted within southern England by a native mosquito. In the United States, with its hot and wet delta cities, the risk of malaria becoming endemic once again is more serious (particularly as mosquitoes have found new places to breed such as the vast dumps of tires filled with stagnant water which defile parts of the American landscape).

The malady is caused by a single-celled parasite, one of several species of *Plasmodium*, which is transmitted by mosquitoes. Mosquito females are deadlier than the males, as they drink blood (which is needed to produce eggs). The parasites are injected from the insect's salivary glands and pass to the recipient's liver. Here, they multiply enormously; one infective cell can divide to produce ten thousand descendants. These enter the blood, break into red cells and divide again. They digest the hemoglobin, multiplying as they go. The *Plasmodium* cells need iron, which they take from the hemoglobin of their host. As a result, to give undernourished African children iron supplements can lead to a new eruption of malaria, which had been lying low. One of the new drugs acts by mopping up iron in the blood and—effectively—starving the parasite.

If the parasites enter the brain there may be a fatal cerebral malaria. Even if they do not, there are bouts of fever as new waves emerge from the reservoir in the liver. Many of malaria's symptoms are due to the release of iron and other toxic breakdown products as the blood is digested.

Once the sufferer has been bitten by a mosquito, the parasites go into their next phase. Within a human being, *Plasmodium* has a life of blameless rectitude, doing little but producing tens of thousands of identical copies of itself. In the mosquito it has a sex life. Males and females mature and mate, producing new genetic combinations among their offspring. The next generation migrate to the salivary glands, where they are ready to be injected into a vulnerable human and the cycle starts again.

There are several species of malaria parasite. They have a surprising evolutionary history. Some of the genes in their cytoplasm are similar to genes found in green plants. Perhaps, in the distant past, their ancestors were related to single-celled plants (possibly those whose modern equivalents cause the "red tides" which kill fish). Although this is rather speculative, there is better evidence about their more recent evolution. The DNA of the most virulent form, *Plasmodium falciparum,* is similar to that of one which infects birds and this is probably where it came from. Other malaria parasites are biologically closer to those which attack apes. They may owe their relative mildness to a long history of evolution in our relatives.

Falciparum malaria needs a fairly dense human population to keep going. It probably began to attack large numbers of people ten thousand years ago, when Africans shifted from hunting on the savannah to farming on the edges of the forests. It will soon be possible to test this idea: fossilized mosquitoes have been found with human blood inside them which might turn out to contain the agents of malaria.

The symptoms can be recognized in writings from ancient Egypt and China. Hippocrates was the first to point to its association with wet places. The swampy area around Rome—the Campagna—was uninhabited for most of its history because of endemic malaria and the disease destroyed the prosperity of the coastal cities of Greater Greece, such as Sybaris and Syracuse. As a result of malaria the fertile Yangtse basin was abandoned for a thousand years. The disease spread over the whole world with the advance of exploration. It was once common in the swamps and fens of East Anglia, much of which was recovered from the North Sea only in historical times. Its major soccer team, Norwich City, wears yellow jerseys because the local population were once called "yellowbellies" after the jaundice caused by chronic malaria. It killed King James the First and Oliver Cromwell; and Sir Walter Raleigh on the scaffold was concerned that his trembling might be interpreted as fear rather than what it was, malaria.

Although hundreds of millions are infected by malaria and millions die from it there seems to be an uneasy coexistence between parasite and host. Evolution has provided dozens of different ways to foil its activities. The way in which humanity has coped with the disease demonstrates the strengths and weaknesses of natural selection better than anything else. All kinds of defenses have appeared, often different ones in different places. Some are effective, some less so; and some impose a terrible cost on those who use them.

To get into the red cell, the parasite must stick to it. It uses the Duffy blood group as the attachment site. In West Africa many people carry a mutation which alters the shape of this site so that the parasite cannot attach to the surface of the cell. They are "Duffy-minus" and are protected against the disease. Duffy and the other blood groups are just a tiny sample of the variable antigens carried by all cells. In malarial Africa there is a combination of these antigens which is rare elsewhere. People with the right set have less severe symptoms should they be infected, perhaps because *Plasmodium* finds it harder to break into their red cells.

One of the great puzzles of biology is to understand why there is so much inherited variation on the surfaces of cells. It is important, as it prevents people from accepting tissues from each other. But it certainly did not evolve to make kidney transplants difficult. Perhaps such diversity is, in part, a relic of a history of natural selection by disease, with particular antigens favored because they protected against specific infections. In the case of malaria, selection must have been quite strong as half of the population of West Africa have protective antigens although the most severe form of the disease has only been around for about five hundred human generations.

Once the *Plasmodium* gets into the red cells there are other defenses. In the peoples of the Mediterranean and the Middle East there is a mutation which reduces the activity of an enzyme inside the cells. This reduces their ability to survive when a parasite gets in, and the cells die, taking the invaders with them.

The most widespread trick which evolution has come up with in its battle against malaria involves changes to the red blood pigment, hemoglobin. There are dozens of such mutations. In some places in West Africa, up to a third of children carry one or two copies of a gene for the mutated hemoglobin known as sickle cell. They have a single alteration in the DNA. This in turn leads to a single change in one of the amino acids, the building blocks which make up the red blood pigment. When a cell from someone carrying sickle cell is attacked by a parasite, the hemoglobin forms fibers and the cell collapses, slowing the growth of the parasite. This is very effective. A child with a single copy of the gene has a ninety percent protection against severe symptoms.

In India and the Middle East there are mutations involving other amino acids in the hemoglobin which act in much the same way— infected cells commit suicide. Italians, Cypriots and others have evolved more drastic defenses. Whole sections of the hemoglobin molecule are deleted. Once again, this slows the growth of the *Plasmodium*. Either of the two chains of amino acids of which hemoglobin is made may be affected. The name of these diseases, the thalassemias, reflects their distribution, meaning, as it does, the anemias of the sea (in this case the Mediterranean). Occasionally, the response to malaria involves the persistence into adult life of a hemoglobin normally found only in the fetus.

The picture looks pretty confused already. Now that it is possible to use DNA to look in more detail at what is going on, things have got even more complicated. What seemed to be the same defense mechanism in separate places turns out to be genetically quite different. There are at least twenty (and perhaps many more) distinct deletions of bits of the hemoglobin chain, and many different protective cell-surface cues. Altogether, hundreds of mutations have been pressed into service in the struggle against malaria. In addition, it looks as if the same mechanism—sickle cell, for example—has turned up independently in populations a long way apart. Four distinct foci of the sickle gene, each associated with a different set of variants in the surrounding DNA, are known from Africa, with another from India.

There are even a few patches of sickle-cell hemoglobin in Europe where in some places the mutation is carried by people with white skins. One of these is in the town of Coruche, in central Portugal, where malaria was once common. Although most of the DNA of these people is similar to that of other Europeans, the DNA around their sickle-cell gene is of a type only found in West Africa. The Portuguese brought home the first slaves from Africa in 1444 and, a century later, the Algarve was almost entirely populated by Africans and their children, a high proportion of whom had a white parent. Many of these children must have carried sickle cell. This protected them against the local disease, malaria, so that this particular African gene flourished and spread although those for black skin were absorbed into the local population and, after hundreds of years, lost from sight.

The malaria story contains some important lessons for the theory of evolution. Its battles are not great set-pieces, like Waterloo or Stalingrad. Instead they are a series of skirmishes. The crowds around the Russian Parliament threw up barricades after the coup against Mikhail Gorbachev. They grabbed whatever was at hand to make a rough and ready barrier which, even if it did not stop the tanks, slowed them down. Natural selection responded to malaria in the same way. Whenever a mutation which might be useful turned up it was used to try and halt the invader. In different places, different genes became available; and the first one at hand was used even if it was not the best. The solution which emerges may be wasteful and inefficient. This ability to "make do and mend" is characteristic of evolution. It explains why no creature is a beautifully economical solution to the problems of its own history and why life is, basically, such a mess.

The evolution of resistance to malaria shows better than anything else the expediency of existence. There is a famous anatomical example of the same thing. In all mammals, one of the cranial nerves takes a slight detour around a vertebra in the neck. In giraffes the neck is enormously extended—but the nerve, far from taking a short cut direct to the brain, goes all the way down to the bottom and back up again. As the genetics of malaria resistance shows, awkward solutions to an evolutionary dilemma are common even at the molecular level. Perhaps they will help to explain (although as yet we have no idea how) why most of the structure of DNA also seems to be, to put it bluntly, a complete shambles.

When faced with an emergency, people often turn to crude solutions which turn out to be expensive in the long run. Evolution does the same thing. Some of the protective mechanisms against malaria damage the populations which take advantage of them. When the sickle-cell mutation first appeared it was rare, so that nearly every copy was partnered by a matching copy of the unchanged gene. This combination protects against infection and leaves its carriers in good general health. As sickle cell became more common, people with two copies of the altered hemoglobin, one from each parent, appeared. They suffer from sickle-cell anemia, a severe disease which can kill. Their red cells collapse even when the parasite has not entered, giving rise to a range of crippling symptoms. These include brain damage, heart failure and paralysis. In some places, around one child in ten is born with this condition.

This is a high price to pay for protection, but is unavoidable once a population starts using this particular gene. Some of the other mechanisms for fighting the disease (including the thalassemias) incur the same cost. As more than one person in twenty worldwide carries one or other of these genes, hundreds of thousands of children with inherited anemias are born each year. Again, this does not add much weight to the idea of natural selection as a benign designer.

Other genetic variants which we now see only as inborn disease may themselves be, like sickle cell, relics of a defense against infection (perhaps against illnesses which have now disappeared). Sickle-cell anemia is found in American blacks, although they are not exposed to malaria. If its association with infection elsewhere in the world was not known its presence in that racial group would be a mystery. Other ethnic groups have their own inborn illnesses. One Ashkenazi Jew in thirty is a carrier of the gene for Tay-Sachs Disease. Those who inherit two copies suffer from an unpleasant and ultimately fatal degeneration of the nervous system. There is some evidence that families carrying this gene had ancestors more resistant to tuberculosis than the average. As TB was common in the European ghettos from whence most of them came, Tay-Sachs may be the relic of a system of protection against infection which looks much like sickle-cell anemia. The cost of past protection is still being paid by their descendants. Other diseases —such as ankylosing spondylitis, or "poker spine"—are more likely to strike people who carry certain cell-surface antigens. Perhaps this too is a relic of natural selection by diseases which have now gone.

Malaria has other attributes which make it an unforgiving enemy even in the face of modern medicine. Many diseases have been beaten by vaccination. By injecting a weakened version of the parasite the body can be persuaded to produce antibodies which will attack the real thing should it invade. The eradication of smallpox is the most spectacular example of the success of this approach. The idea of a malaria vaccine has proved a will-of-the-wisp. New work on the genetics of the parasite shows how difficult the task will be. *Plasmodium* is enormously variable. One of its many surface antigens (which would have to be mimicked by any successful vaccine) exists in forty different forms. Seven have been found in just one Sudanese village. The parasite's sex life only makes things worse. There are several genes, scattered all over its fourteen chromosomes, which produce its own cell-surface antigens. Every time *Plasmodium* gets into a mosquito and has sex, they are reshuffled into new and unique combinations. Many patients with malaria are infected with more than one strain of the parasite, so that new mixtures of antigens appear all the time. It will be many years—if ever—before malaria goes the same way as smallpox. The disease has turned out to be a subtle and effective opponent. The genes which protect against it will be needed for a long time yet.

Nevertheless, plenty of infections have gone forever. Some optimists claim that the conquest of disease, cold and starvation means that natural selection has come to an end. If there is one general rule to be learned from evolution, it is to expect the unexpected. It is quite possible that new pestilences will appear and cause just as much damage as malaria, or that those which are apparently near extinction will stage a resurgence, as has malaria itself.

The history of the battle against disease says some useful things about evolution. Far from designing a simple and effective protection, whenever a straw appeared natural selection clutched at it. Selection acted like a handyman rather than a craftsman. Its products often seem to be amazingly badly—not to say extravagantly—planned and roughly made. If man is indeed made in God's image, it does not say much for divine engineering.

This haphazard approach has its strengths. When used by engineers and computer programmers it can provide subtle and unexpected ways of dealing with problems. Turbine blades, spray nozzles and works of computer art have been developed using the logic of selection to do what it does in the living world: to produce a complicated design without a designer.

However, natural selection has never in its three-billion-year history produced a turbine blade or even a wheel, let alone a work of art, although it has managed to generate eyes, brains and other organs of great complexity. This is because of selection's greatest weakness, its plodding approach. A wheel, or a watch, needs some long-term ideas. There is no such thing as a first tentative step and to make either of them demands an intellectual leap which selection cannot make. Natural selection has superb tactics, but no strategy—but tactics, if pursued without thought for the cost and for long enough, can get to places which no strategist would dream of.

12 | Caliban's Revenge

THE PLOT of the Victorian novel *Daniel Deronda*, by George Eliot, is a convoluted one. It revolves around the adventures of Daniel himself, the adopted son of a baronet. After some hundreds of pages he develops a surprising interest in things Hebrew and—some time later—it transpires that Daniel Deronda was, without knowing it, the son of a Jewish woman. His biology had triumphed over his upbringing.

For many people the role of inheritance compared to that of experience is an obsession. It is an obsession that goes back long before genetics. Even Shakespeare had a say: in *The Tempest* Prospero describes Caliban as "A devil, a born devil, on whose nature Nurture can never stick." There are still endless discussions about whether musicality, criminality or—most fashionable of all—intelligence is inherited or acquired. More seriously, there is debate about the role of genes and environment in controlling illnesses such as cancer or heart disease.

Galton, in *Hereditary Genius,* went to great lengths to show that talent runs in families and was coded into our biology. Oddly enough, he never pointed out that more than half his "geniuses" turned up in families with no history of distinction at all. Such was the power of prejudice that he concentrated only on those who supported his hereditarian views. Most claims that talent (or, for that matter, lack of it) is inherited are based, like Galton's geniuses, on little more than a series of selected anecdotes. Even the descendants of Johann Sebastian Bach disappeared from the musical firmament after a few generations. I sometimes illustrate the ineffectiveness of using family similarity to establish the importance of biology by asking what, of all our attributes, is most similar among British or American parents and their children—or their sisters and their cousins and their aunts. The answer is bank-balance.

Nevertheless, the question of nature versus nurture is of endless fascination. There are dozens of studies which purport to show that behavior is under genetic control. Particularly in America whole sets of degenerate families were once held up for inspection: the Tribe of Ishmael, the Jukes Clan and the Kalikaks (whose pseudonym is Greek doggerel for good/bad). One was traced to an eighteenth-century sailor who married an upright woman but had an affair with a slattern. His wife's branch gave rise to a family of spotless virtue while the other was a burden on society. This proved, needless to say, that morality lies in the genes.

Most modern geneticists find queries about the relative importance of nature and nurture in controlling the normal range of human behavior dull, for two reasons. First, they scarcely understand the inheritance of complex characters (those like height, weight or behavior which are measured rather than counted) even in simple creatures like flies or mice and even when studying traits like size or weight which are easy to define. Second, and more important, geneticists know that the perpetual interrogation—nature *or* nurture?—is largely meaningless. Its only answer is usually that there is no valid question.

Although genetics is all about inheritance, inheritance is certainly not all about genetics. Nearly all inherited characters more complicated than a single change in the DNA involve gene and environment acting together. It is impossible to sort them into convenient compartments. An attribute such as intelligence is often seen as a cake which can be sliced into so much "gene" and so much "environment." In fact, the two are so closely blended that trying to separate them is more like trying to unbake the cake. Failure to understand this simple biological fact leads to confusion and worse.

Not far from Herbert Spencer's (and his neighbor Karl Marx's) tomb, in Hampstead—a notably affluent part of London—is a large red-brick house. It was occupied by Sigmund Freud after he fled Austria to avoid racial policies which descended from the Galtonian ideal. On his desk is a collection of stone axes and ancient figurines. Freud's interest in these lay in his belief that behavior is controlled by biological history. Everyone, he thought, recapitulates during childhood the phases which humans experienced during evolution. Freud saw unhappiness as a sort of living fossil, the emergence of ancient behavior which was inappropriate today. Like Galton he viewed the human condition as formed by inheritance. The libido and the ego are, he wrote, "at bottom heritages, abbreviated recapitulations of the development which all mankind has passed through from its primaeval days." Freud hoped that once he had uncovered the inherited fault which underlies mental illness, he might be able to cure it.

Today's Freudians have moved away from their master's Galtonizing of behavior. They feel that nurture is more important. Analysis looks for childhood events rather than race-memories. In so doing it is in as much danger as was Freud of trying to unbake the cake of human nature. Any attempt to do so is likely to prove futile.

The Siamese cat shows how futile the task may be. Siamese have black fur on the tips of the ears, the tail and the feet, but are white or light brown elsewhere. The cats carry the "Himalayan" mutation, which is also found in rabbits and guinea pigs (but not, unfortunately, in humans). Breeding experiments show that a single gene inherited according to Mendel's laws is involved. At first sight, then, the Siamese cat's fur is set in its nature: if coat color is controlled by just one gene then surely there is no room for nurture to play a part.

However, the Himalayan mutation is odd. The damaged gene cannot produce pigment at normal body temperature but works perfectly if it is kept cool. This is why the colder parts of the cat's body, its ears, nose and tail (and, for a male, its testicles) are darker than the rest. An unusually dark cat can be produced by keeping a typical Siamese in the cold and a light one by bringing it up in a warm room. Inside every Siamese is a black cat struggling to get out. It is meaningless to ask whether its pattern is due to gene or to environment. It is due to both. What the Siamese cat—and every living creature—inherits is an ability to respond to the environment in which it is placed.

Some inborn diseases show this effect particularly clearly. The recessive abnormality phenylketonuria (or PKU) affects around two hundred American children a year. They have an inherited defect in a particular enzyme which makes it impossible to process an amino acid, phenylalanine, which is found in most foods. As a result they build up large amounts of a harmful by-product. Untreated, the children have low intelligence and die young. The fate of those with PKU is, it seems, sealed by their genes.

But most PKU children born today behave normally and lead ordinary lives. A change in the environment saves them. If they are diagnosed early (and most babies are now tested at birth), they can be given a diet which lacks all but a tiny amount of phenylalanine. They then develop as healthy children. Their nature has been determined by careful nurturing and there is no answer to the question of whether their gene or their environment is more important to their well-being.

Hundreds of human variants show the same interaction between heredity and environment. Many are medically important. There is now a whole science of "pharmacogenetics" which depends on individual differences in the response to drugs—tobacco and alcohol included. The very existence of the genes involved was unknown until humans began to manipulate their own circumstances by chemical means. A few people carry an inherited variant which makes them fatally sensitive to a muscle relaxant used before surgery; and everyone is now tested to see whether they are at risk before the drug is given. There are even inherited differences in the ability to cope with food. One of the stranger injunctions of Pythagoras was a warning to his followers not to eat broad beans. He died because, pursued by a mob enraged by his philosophical views, he refused to escape across a beanfield. Pythagoras lived in the Italian city of Croton. Many of its modern inhabitants feel unwell if they eat partly cooked beans. They carry a gene for thalassemia (which protects against malaria). One of its side effects is to remove the ability to break down a chemical found in broad beans (and another one used as an anti-malarial drug). When gene and bean (or drug) are brought together, the results can be unpleasant or, in the case of the drug, worse.

All this means that the boundaries between inherited disease and what was traditionally considered to be governed by environment are becoming blurred. It is changing the whole way we think about medicine. Individual treatments may soon be tailored to a patient's biological heritage. Doctors will be able to protect those carrying a particular variant from the circumstances which threaten them. Two birth disorders, anencephaly and spina bifida, cause the spinal cord to develop incorrectly. Family studies suggest that they are inherited. Their incidence shot up in Holland after the famine of 1945. Part of the problem has to do with poor diet. Both are frequent in Northern Ireland and in Scotland (places with notoriously unhealthy eating habits). Mothers who have had an affected child now take vitamin supplements during later pregnancies. This reduces the chance of their genes damaging their children.

The term "cancer" covers a multitude of conditions. All are due to a failure to control cell division. Cancer genetics is now a field large enough to deserve a book to itself. We do understand enough about the disease to see that cancer is a Siamese cat of an illness: very often, the chances of contracting it depend both on the genes one has and the circumstances with which they are faced.

Some cancers are more common among those exposed to a particular hazard. Many chimney-sweeps—the heroes of Charles Kingsley's fantasy *The Water Babies*—died of skin cancer, which appeared first on the scrotum. The English physician Percival Pott suggested that soot was to blame. He was right. Soot, oil and tar are now known to contain many carcinogenic chemicals. Other cancers (such as retinoblastoma, a degenerative disease of the retina) run strongly in families, with no obvious environmental link.

The causes of cancer run all the way from gene (which is predominant in retinoblastoma) to environment (which plays an important part in scrotal cancer) usually including both. Workers in the early oil industry believed that fair-haired people with freckles should not be employed as they were more at risk of "sootwort," as scrotal cancer was known. As such people are more likely to get skin cancer when exposed to sunlight, there may be some truth in this. Even lung cancer has a genetic component. It is, as everyone knows, most common among smokers; and tobacco smoke contains many of the carcinogens to which the sweeps were exposed. However, smokers who are unfortunate enough to inherit a gene for susceptibility are far more likely to contract cancer than are those who do not. If everybody smoked, lung cancer would be a genetic disease.

Diet, too, may be important. Cancer of the liver is common in parts of China and in southern Africa. The local food is contaminated with aflatoxin, a chemical produced by mold. Those with the disease all have a new mutation in a gene whose normal role is to prevent cells from dividing uncontrollably. The mutation is of exactly the kind produced by aflatoxin in the laboratory and the peoples of these regions have high levels of the poison in their blood. The case seems watertight. By improving food storage and controlling mold, liver cancer should be easy to control. Unfortunately, poverty means that this simple environmental change is unlikely to be achieved.

Many cancers are sparked off by a mutation, perhaps because of exposure to an environmental mutagen such as a chemical or radiation. Every cell contains a hundred or so "proto-oncogenes" involved in the control of cell division and in the interactions of cells with each other.

Sometimes, these mutate. They give rise to oncogenes, which are implicated in the early stages of certain cancers. There may also be mutations in "tumor suppressor genes," whose normal role is to prevent cells from dividing uncontrollably.

Some people inherit a mutant form of a gene for cancer susceptibility from their parents. They are likely to develop a specific cancer or—sometimes—one of several forms of the disease. Others inherit genes for cancer susceptibility which are unduly liable to mutate. They are more at risk from carcinogenic chemicals than is the general population. The picture is further complicated by the fact that viruses, too, play a part. They can carry oncogenes, which insert themselves into the host DNA.

The alloy of gene and environment needed to produce a cancer is seen at its clearest in Africa. Burkitt's lymphoma is a tumor of the lymph nodes, which usually begins in the neck and jaw. It is common in East Africa. A virus—the Epstein-Barr virus—is involved. Any infection by the virus provokes an immune response. In Africa, the immune system is often already under strain because of chronic infection by malaria. Among some of the cells responsible for making antibodies against the parasite there may be a genetic change. A chromosomal mutation fuses an antibody gene to a particular oncogene. This sets the cells on course for cancer once there is further strain on the immune system after Epstein-Barr virus attacks. Only in places where the virus infects people inherently predisposed to respond to it—and already weakened by infection—is Burkitt's lymphoma common.

Because cancer involves such an intimate interaction between internal and external causes moves are afoot to try to protect those at risk. Heavy smokers—about one in ten of whom will develop lung cancer—are given vitamin A in the hope of reducing the effects of mutations in their lung cells. Those who inherit a gene predisposing them to colon cancer are treated with aspirin before they develop symptoms as this might reduce its damaging effects. Cancer is sometimes seen as a kind of biologically programmed Nemesis about which nothing can be done. An appreciation of the importance of the environment (which can often be changed) gives new hope. Sometimes, this raises distressing problems of its own. A small proportion of the women who develop breast cancer do so because they have a gene which predisposes them to the disease. Now that the gene can be identified, they are faced with the prospect of having a breast removed before any symptoms appear.

A harmful gene can sometimes become obvious only when the environment changes. Diabetes results from a failure to maintain the correct level of sugar in the blood. Some patients show the disease early in life, others later. Although some diabetics can successfully be treated with insulin there is another form of the disease, non-insulin dependent diabetes, which is more common. Diabetics can suffer from a variety of symptoms. They include heart disease, gangrene, kidney failure and blindness. In parts of the third world non-insulin dependent diabetes has become an epidemic.

The people of the Pacific island of Nauru have had riches thrust upon them because of phosphate mining. Instead of fish and vegetables they eat fat and sugar. Eight out of ten adult Nauruans are diabetics and the island now has one of the shortest life spans in the world. Perhaps the local genes for susceptibility to sugar were at an advantage in times when starvation was followed by glut. Only when glut became the norm did they become dangerous. A change in the environment has led to genetic disease.

American Indians and the peoples of the Pacific both came from Asia a few thousand years ago. The biological heritage of the diabetics of Nauru is present in the natives of the New World. Many Mexican-Americans suffer from "New World Syndrome"; they are diabetic and may be very fat. The disease is rare among white Americans. The risk of getting it goes up with the number of Amerindian ancestors an American citizen has—at first sight a good case for saying that genes are of paramount importance. But the disease is almost unknown among American Indians living in their home communities, in Mexico or further south. It affects them only if they change their diet by moving to the United States. Differences among Americans in the incidence of New World Syndrome arise from both nature and nurture.

To many in the third world, sugar is a toxic chemical. Other chemicals, such as those used in industry, are even more dangerous; and there exists—as for sugar—a set of inherited mechanisms which can make them safe. There is variation in how active these genes are, and the World Health Organization has suggested that employers screen workers and advise the susceptible to work elsewhere. There is nothing new in this. After all, those who are red-green color-blind already know that they cannot be train drivers. There is, however, the possibility of using genetics as an excuse *not* to improve the environment. The journal *Chemical Week* once wrote that ". . . it makes no economic sense to spend millions of dollars to tighten up a process which is dangerous only for a tiny fraction of employees . . . if the susceptible individuals can be identified and isolated from it."

The new understanding of the way in which DNA interacts with its surroundings is, if nothing else, likely to change attitudes to risk. Most people know that smoking causes cancer and that a fatty diet may lead to heart disease. Certain genes predispose their carriers to the harmful effects of smoke or fat; and some individuals may be able to drink, smoke or eat lard with impunity. Perhaps it will become possible to choose the vices best fitted to ourselves. Soon, people may be forced to change their perceptions of personal danger. Propaganda about smoking and lung cancer has not been particularly effective. Those exposed to it have an infinite capacity to assume that if only one smoker in ten contracts the disease, then that smoker will be someone else. When molecular biology makes it possible to identify exactly who will get cancer if they smoke, individual terror may prove to be a better deterrent than is collective risk.

Diseases with a simple pattern of inheritance—like PKU—are not individually common. More frequent disorders such as cancer and heart disease certainly run in families, but their inheritance is harder to study. Many different genes are involved and the environment plays a major part. One way of exploring them is to use twins, nature's own experiment in human genetics.

Twins are of two kinds, identical and non-identical. Non-identical (or fraternal) twins come from the fertilization of two eggs by two sperm. There have even been a few cases in which they turn out to have different fathers. Such twins have half their genes in common and are no more similar than are other brothers or sisters. Their situation is —yet again—described in that fount of early genetics, the Old Testament. Jacob and Esau were twins; but "Esau was a cunning hunter, a man of the fields; and Jacob was a plain man, dwelling in tents." They looked quite different—"Behold, Esau my brother is an hairy man, and I am a smooth man"—and even have different ways of speaking: "The voice is Jacob's voice, but the hands are the hands of Esau."

Fraternal twins are not rare. In marmoset monkeys most births are of this kind. For no obvious reason, the numbers of such twins vary from place to place. In Europe (and in Americans of European ancestry), about eight births per thousand are of fraternal twins (France having rather fewer and Spain rather more than the average). Among the Yoruba, in Nigeria, the figure is five times higher. Older mothers tend to have more twins, as do those who have already had several children.

Their numbers have been dropping in industrialized countries for the past century.

Identical twins are rarer, at about four per thousand births, a rate which does not change much in space or time. There are few mammals in which they are common, but the armadillo nearly always gives birth to identical quadruplets. Identical twins result from the splitting of an egg which has already been fertilized. They share all their genes. Such twins have long been a source of legend. Castor and Pollux, the heavenly twins, were identical; as were their equivalents in Germanic legend, Baldur and Hodur (not to speak of Romulus and Remus, the founders of Rome).

There are several ways in which twins could be used to study nature and nurture. Apparently the simplest (but by far the least common) is to find identical twins separated at birth and brought up in different households. If a character is under genetic control the twins should stay the same in spite of their differing circumstances. If the environment is more important, each twin should grow to resemble the family with which they spent their childhood.

This simple plot is the basis of a great deal of fiction, in science as much as in literature. In the early days, many studies claimed to show that identical twins reared apart were similar in size, weight or sexual orientation. Much of this work had problems. Often, the adoptive families were similar in class and social position. Sometimes, the twins even knew each other as they grew up. Twins who believed themselves to be identical turned out to be fraternal when blood tests were used. Even worse, there have been persistent accusations of fraud, particularly where intelligence testing is involved. All this means that most of the older research on twins reared apart has been discarded. However, there is the beginning of a new study of this kind which shows quite convincingly that some traits of personality—aggressiveness, introversion and so on—have a genetic component. This does not, of course, imply that nurture can be disregarded. An intrinsically aggressive man may be calm until he is given a chance to express his genotype by joining the army.

There is a more subtle way of using twins. It involves comparing the similarity of identical twins with that of fraternals. The argument is that, as both kinds of twins are brought up within their own family, the degree to which they share an environment is the same. Any greater resemblance of identical twins to each other must show that genes are involved. This approach could be powerful but has its own problems, particularly where studies of behavior are involved. Although both types of twin are brought up together, identicals may copy each other's behavior on purpose, making them appear more similar for reasons quite unconnected with their biology. The very fact of being identical twins—perhaps sharing very similar names and dressed in identical clothes—may predispose to mental disease. Twins often have a poor environment before birth as they share a placenta. This happens more in identicals so that their similarity may be due more to a shared environment than at first appears.

Nevertheless, this approach has had its successes. Members of a pair of identical twins are twice as likely to suffer from coronary heart disease than are those of a pair of fraternal twins; and five times as likely to have diabetes. Even tuberculosis is more commonly shared between identical than fraternal twins, suggesting that there may be an inherited basis for susceptibility. Other characters, such as schizophrenia or the age when a baby first sits up, are more similar for identical than for fraternal twins.

The argument about nature and nurture is of more than merely scientific or medical interest. It has been rehearsed endlessly by those with one or the other political ax to grind. As much of early genetics used an ax sharpened in the fires of Social Darwinism it has often been used as to justify injustice. Not much has changed since the Social Darwinists. As genetics has hit the headlines, there has been a new acceptance of biological theories of human behavior. Arson, traditionalism and even zest for life have all been blamed on the DNA. The 1960s were the decade of caring. A child's inability to concentrate was blamed on poor teachers. Then, in the boom years which followed, there was the "working-mother syndrome" in which a parent's absence was held to be at fault. Now psychologists have invented "attention-deficit disorder": something intrinsic to the child and coded in the genes. Psychology's obsessive need to dissect biology from experience is alive and well. A Harvard professor, no less, has found that students with hay fever appear to be unusually shy. This proves, he thinks, that "there is a small group of people who inherit a set of genes that predispose them to hay fever and shyness." Those interested in the genetics of human behavior do not themselves suffer from undue modesty. There have been announcements of the discovery of single genes for manic depression, schizophrenia and alcoholism. All have been withdrawn.

Nevertheless, the biologizing of behavior goes on apace. Some of it may even be justified. Crime is largely a male preserve, and in the broadest sense can be tracked down to the single short DNA sequence on the Y chromosome that causes a fetus to develop into a male. The link between gene and crime is such a distant one as almost to lack meaning; and most males, of course, are not criminal at all.

There have been many attempts to associate antisocial behavior with more specific genes. Some legal textbooks have sections dealing with how to predict criminality on biological grounds and brain scans are admitted as evidence in U.S. courts. The next step—geneticizing crime: "It wasn't me that did it, it was my genes"—cannot be far behind. A tenuous claim that an extra Y chromosome made men more violent led to the suggestion that babies should be tested and watched for criminal tendencies. In this brave new deterministic world, the eugenical cranks who believed that antisocial behavior was coded in the genes would find themselves quite at home.

None of this means, of course, that genes and crime can have nothing to do with each other. There has recently been discovered a Dutch family in which several of the men have carried out violent assaults. Most of those who came to the attention of the law shared a variant form of a gene whose product is involved in the transmission of impulses between nerve cells. Interfering with nerve transmission using drugs is well known to alter behavior. Faced with circumstances which would not perturb most men, these males fly into a rage. It is not yet known how common the variant is among law-abiding members of the community.

There is one form of behavior, shared to some extent by most people, in which the relative importance of nature and nurture continues to raise passions. Homosexual attraction is almost universal at some time in everyone's existence. Some people continue to prefer their own sex throughout their lives. Exclusive homosexuality is a convenient subject of study for those interested in the genetics of human conduct: unlike, say, musicality or schizophrenia it is easy to identify, relatively common, and—at least in the United States—no longer much concealed.

A study of a group of American male homosexuals gives a hint of an association between such behavior and a group of genes on the X chromosome. First, the brothers of gay men were more likely to be gay than are males in the general population. This does not in itself say much, as brothers share an environment as well as genes. However, gay men's relatives on the mother's side were more liable to be gay than were those on the father's side, suggesting that the trait is passed through females in the family. Again, this is not in itself proof of an innate predisposition (although it hints at a gene on the X chromosome). The best evidence came from looking at the X chromosomes of pairs of homosexual brothers. Most shared a particular segment of DNA toward one end of that chromosome. Somewhere in its hundred or more genes may be one that inclines some of its carriers to homosexuality. Unlike the variant nerve transmitter in the Dutch study, there is no indication at all of what that gene might do and—just as for violence —there are certainly many homosexuals who do not have it. If the research had been done on a less controversial character most geneticists would accept that the first step to tracking the gene down has been taken. More work is needed before the result is confirmed.

The response by some—but not all—of the gay community was surprising. Many were happy to use biology as a justification for their way of life. The idea that sexual preference was inherited meant, it seemed, that homosexuality was not contagious and that attempts by bigots to dismiss (for example) homosexual teachers were not justified. More important, it gave a welcome sense of separation: of difference for reasons beyond individual control.

All this is disconcerting to biologists, many of whom spent years fighting the idea that crime, poverty or behavior are inborn and cannot be altered by social means. There is a grim Calvinism about the idea that talent, mental health or sexual preference are writ in DNA. But, for some members of the gay community, the idea seems to be more easy to accept than is that of unfettered choice.

In fact, this new hereditarian orthodoxy is not very far from the old liberalism. Both ask too much of biology. The recent debate echoes an almost forgotten dispute of the 1930s. The German geneticist Theobald Lang claimed to have found that the sisters of homosexual men had somewhat masculine characters, and that male homosexuality might therefore be inherited. Whatever the accuracy of his claim (and there is no indication from more recent work that it is correct) that early hint of a "gay gene" gave rise to two absolutely opposed responses. The Nazis, needless to say, took the brutal view: "they are not poor sick people to be treated; they are enemies of the state to be eliminated!" In contrast—and faced with exactly the same information— the German socialist medical association (then in exile) wrote that as "homosexuality is inborn and not subject to the free will of the individuals who come into the world with this inversion" then the laws forbidding it should be abolished.

Like some members of today's gay community, both left and right in Germany felt that if homosexuality was innate it must be outside the control of homosexuals themselves. Each political group saw its response—eugenic sterilization versus legalization—as consistent and logical. Neither asked what is meant by a gene "for" something—homosexuality, intelligence or criminality—although, as we have seen, that is by no means a simple question. If anything, the story of the German "gay gene" points up the irrelevance of genetics to political opinion: whatever genetic basis a character may have, preconceived views about its merits are unlikely to be changed by science.

Nowhere has the difficulty of separating science from politics, and the confusion about nature and nurture, had a more malign effect than in the study of differences among human groups. If one reads older textbooks on race—I have, and it is a depressing experience—sooner or later one comes to the question, always treated with a certain prurience, of inherited differences in intelligence. In the early days that such differences existed and that they were inborn seemed blindingly obvious. Linnaeus himself classified humans as *Homo sapiens,* thinking man. For the species as a whole, he could be no more precise in his definition than *"Homo, nosce te ipsum"*: Man, know thyself. His description of the different varieties of humankind, however, used behavior as an important character. Linnaeus' definition of an Asian, for example, was someone who was yellow, melancholic and flexible. Even forty years ago, racial stereotypes of the most predictable kind were the norm among psychologists.

Much of the work on inherited differences in intellect between races is contemptible and most of the rest is wrong. The wrong argument usually goes like this. Blacks do less well than whites on IQ tests, so that they are less intelligent. The IQ scores of parents and children are similar, so that intelligence is controlled by genes. The difference between blacks and whites must therefore be genetic.

This argument is deceptively simple. It was once used in the U.S.A. as an excuse not to spend money on black education, and a variant of the theory, which sees working-class rather than black children as victims of their genes, is often employed in Britain by those who resent spending money on state education. Simple though it may be, it is utterly false.

I have no idea whether IQ tests are an unbiased measure of intelligence; what they measure is, I hope, known to those who design them. The similarity of parents and children in the ability to do the test does not in itself tell us much, as families share the same environment as well as the same genes. New twin studies do suggest that there is an inherited component in IQ. Some claim that as much as seventy percent of the variation in IQ score within a population is due to variation in its genes. This figure seems high, but can be accepted for the present. At first sight it looks like powerful evidence for the view that any racial differences in IQ must be biologically programmed.

In fact it has no relevance to understanding whether differences in intelligence—if they exist—between races are inborn or acquired. The reason why can be seen by looking at some other characters which show racial differences. In the United States, the blood pressure of middle-aged black men is about fifteen percent higher than that of whites. Twin studies show that about more than half the variation in blood pressure within a group is due to genetic variation. The figures for the blood pressure story look remarkably like those for IQ, although in this case it is blacks who come out with a higher score.

Doctors and educationists have a subtle difference in world view when faced with figures like these. Doctors are optimists. They concentrate on the environment, the fact that blacks smoke more and have poorer diets than whites, and try to change it. In the U.S.A., optimism paid off. High blood pressure and heart disease among blacks is less of a problem than it was.

Many educationists are less hopeful. To them, the existence of inherited variation in intelligence means that there is no point in trying to improve things by changing the environment. Blacks, they say, have worse genes. These cannot be altered, so that it is futile to spend money on better schools. In some countries their theory has been proved wrong. Over the past twenty years the average IQ score of Japanese children has risen to more than ten points higher than that of Americans. Not even the most radical hereditarian claims that this is due to a sudden burst of genetic change in Japan. Instead, the schools are getting better.

Both genetical and environmentalist views of blood pressure or IQ are naive. Characters like these are shaped by both gene and environment so that it is meaningless to ask about genetic differences except in populations living in the same conditions. I once did a simple experiment with a group of students. I divided them on the basis of hair color. The fair-haired group were sent downstairs for coffee. The other set measured their own resting blood pressure. I then summoned the coffee drinkers, who took their blood pressures. Not surprisingly, as they had just run upstairs and drunk coffee their average score was higher than that of the dark-haired students. There was an association between blood pressure and hair color.

To many of the students this made it obvious that there was a genetic difference in blood pressure between dark- and fair-haired people. Only when let into the simple secret of the differences in exercise and caffeine consumption between the groups was it obvious what is wrong. The students had made the same mistake as the educationists. High heritability of a character combined with a difference in its value between groups need not say anything about genes. The race and IQ story is largely one of a dismal failure to understand basic biology.

A belief in heredity, rather like a belief in predestination, is a good excuse for doing nothing. At least the environmentalist version can be used to try to improve things. The genetic view is usually a chance of blaming the victim, a way of excusing injustice because it is determined by nature. In the last chapter of *Daniel Deronda*, biology wins: the hero's fate is coded in his genes. He returns to his ancestral roots and marries Mirah Mordecai, with the Cohen family in attendance. His admirer Gwendolen Harbeth is left to console herself with the memory of her unlikeable mate Henleigh Grandcourt, horribly drowned a few pages earlier. Determinism triumphs, which is convenient for the novelist. Unfortunately—or perhaps fortunately—real life is a lot more complicated than that.

13 | Cousins Under the Skin

NINETEEN HUNDRED AND SIX was a successful year for the Bronx Zoo. A new exhibit was pulling in the crowds. An African Pygmy —Ota Benga by name—was in the same cage as an orangutan. The exhibit caused an uproar, not because it was a degrading spectacle, but because it promoted the idea of evolution, that apes and humans were related. After a time, Ota Benga was released—partly because of his habit of shooting arrows at those who mocked him. He moved to Virginia, where he committed suicide a few years later.

The Bronx Zoo view of human evolution was once widespread. Linnaeus himself, who first classified animals and plants, put the idea neatly in 1754: "All living things, plants, animals and even mankind themselves, form one chain of universal being from the beginning to the end of the world." The Great Chain of Being sees evolution as a smooth progress, a seamless transition from the primeval slime to the Clinton administration. Linnaeus named our own species as *Homo sapiens*—and recognized several distinct varieties. As well as the yellow, melancholic and flexible *asiaticus* there was *europaeus*, white, ruddy and muscular; *americanus*, red, choleric and erect; and *afer*, black, phlegmatic and indolent.

The different groups of humanity were at different stages. Africans were at the bottom, close to the apes, Asians somewhere in between, and white Europeans—of course—at the top. Victorian writers did not hesitate to put the idea bluntly. Robert Chambers, who wrote an influential book on evolution fifteen years before Darwin, claimed that

"Our brain . . . passes through the characters in which it appears in the Negro, Malay, American and Mongolian nations, and finally is Caucasian. The leading characters, in short, of the various races of mankind, are simply representatives of particular stages in the development of the highest or Caucasian type . . . The Mongolian is an arrested infant, newly born.''

The theory that races are biologically different has a long and ignoble history which has brought misery and death in its wake. It reached into medicine. Most people have seen children with Down's Syndrome, which is due to an error in their chromosomes. This was called by its discoverer, Langdon Down, "Mongolism" in 1866 for what seemed to him a good scientific reason—these children had slipped a couple of rungs down the evolutionary ladder to resemble a lower form of human life, the Mongols. Oddly enough, a Japanese friend once told me that in his country the same condition is called Englishism. The idea now seems ridiculous and we know that Down's Syndrome is due to an error in the transmission of a particular chromosome which is found in all the groups of humankind and even in chimpanzees.

This chapter is about what biology can and cannot tell us about the differences among the peoples of the world. The history of race illustrates, more than anything else, the limitations of biology in understanding human affairs. Biologists have been talking—or shouting—about race for years. Ignorance and confidence have often gone together. Although politicians usually take scientists a lot less seriously than scientists do, the story of scientific racism, as it was once known, is a grim one. Even when stripped of the bigotry of the past the question of the nature and extent of the inherited differences among races causes controversy today.

I have always felt a certain compassion for those whose ability to despise their fellow men is limited by the color of their victim's skin. It seems to me that genetics has—and should have—nothing to do with judging the value of one's fellow beings. In this sense, the biology of human race has nothing to do with the issue of racism. Modern genetics does in fact show that there are no separate groups within humanity (although there are noticeable differences among the peoples of the world). This is reassuring, but should be irrelevant to the social or political nature of race. To depend on DNA to define morals is a dangerous thing to do. Science evolves. We learn more, and theories alter. This is as true for genetics as for anything else. Views on human biology have changed and may change again. Surely the same should not be true of attitudes to human rights. Where biology stops and principles begin should never be forgotten.

Humanity can be divided into groups in many ways; by culture, by language and by race—which usually means by skin color. Each division depends to some extent on prejudice and, because they do not overlap, can lead to confusion. In 1987, a secretary from Virginia sued her employer for discriminating against her as she was black. She lost the case on the grounds that, as she had red hair, she must be white. She then worked for a black employer and, undaunted by her earlier experience, sued him for discriminating against her as she was white. She lost again; the court found that she could not be white as she had been to a black school.

Nations, too, differ in how they define their racial identity. In South Africa a single African ancestor, however long ago, meant ejection from the white race. In Haiti, Papa Doc proudly proclaimed his nation to be a white one as nearly everyone, however dark their skin, had a white ancestor somewhere. Other countries developed fine distinctions based on skin color. In Latin America two centuries after the invasion of the Spaniards more than twenty races were recognized. The offspring of a Spaniard and an Indian was a mestizo, that of a mestizo and a Spaniard a castizo, a Spaniard and a negro a mulatto, a mulatto and a Spaniard a morisco, a morisco and a Spaniard an albino, an albino and a Spaniard a torna atras and so on in a lengthy and hair-splitting series. All this shows how difficult it is to make an objective definition of what is meant by race.

Races were once supposed to be distinct because they descended from different ancestors. Ham, Shem and Japhet, the sons of Noah, were popular candidates. Anthropology began with the search for perfect examples of each lineage, for racial *types*. Africans, whites and Asians were thought to be separate units, quite different from each other. Perhaps, the early anthropologists thought, in the dawn of humankind every race was a pure and unpolluted line, living in its ancestral homeland. Only in modern times was this purity sullied by interbreeding. Race mixing was hence against nature and to be avoided. Exceptions might be allowed in emergency, as when Saints Cosima and Damian, with the help of divine intervention, transplanted a black leg onto a white patient.

If, anthropologists reasoned, the peoples of today are a confused mix made up of what was once a series of pure races, it might still be possible to identify individuals who represent perfect specimens of one or other of the original groups. Their subject went around in circles for much of its history in trying to sort out divisions into which people could be classified. Its early days were spent in a useless search for the homelands and migration routes of a series of imaginary pure races which were thought to have mixed to produce modern humans. Typological thinking went to absurd lengths. Harvard University was a center of the search for the archetype. Two suitably discreet nude statues stood until recently in the Peabody Museum of Anthropology. They were based on measurements made in the 1930s on dozens of male and female students. Average these out, the argument went, and one would produce an image of the ideal Harvard undergraduate—the highest form of human being. One remnant of this philosophy is the Miss World Contest whose judges—like the early students of human evolution—desperately try, and fail, to find an objective definition of the perfect woman.

Racial types were usually identified from skulls. The word "Caucasian" reflects a claim that the skull which best represented white-skinned people came from the Caucasus Mountains so that—perhaps—the white race had spread from these remote fastnesses. Years were wasted in measuring skulls rather than thinking about evolution. The most popular (and the easiest) yardstick was the cephalic index, the ratio of the length and breadth of the skull. Tens of thousands of skulls from different parts of the world were measured in an attempt to sort out their ancestral stocks.

The work was futile. There is no evidence at all that there are, or ever have been, populations containing individuals all of whom share the same cephalic index. More disturbing for the craniometers, the skull shape of the children of immigrants to America shifted away from that of their parents toward that of the people already there. The shape of skull is in any case affected by natural selection. Populations living in hot places as far apart as Africa and Malaya have similar skull form, which differs from that of Scandinavians or Eskimos. Even if they have different ancestry, they have converged to roughly the same shape. Shared skull form need not mean a common homeland. A little bit of natural selection has eliminated a great deal of history.

So obvious seemed the differences between groups that they blinded scientists to their own results. Samuel George Morton in his *Crania Americana* of 1830 measured two hundred and fifty skulls. The differences were, he thought, clear: Caucasians had larger brain cases than Mongolians and Malays, who in their turn had larger brains than

Africans. A hundred and fifty years later, the same skulls were remeasured with modern instruments. The differences largely disappeared; Morton's effects had been due to omission of some groups—such as Peruvians—which did not fit his ideas, confusion of males and females, and a failure to correct skull size for differences in body size.

In spite of these problems (which led to the abandonment of the cephalic index) early workers had enormous confidence in its value. Skull measurements were used by the Nazis in an attempt to sort out those with Jewish ancestry. The Frenchman Georges Vacher de Lapouge who wrote in 1887 "I am convinced that in the next century millions will cut each others' throats because of one or two degrees more or less of cephalic index" was more correct than he feared.

Races could also be classified by language. The term "Aryan," which gained such sinister overtones, came originally from the idea of a talented people, the Arya, who migrated from a homeland somewhere in the east, bringing their inheritance and their language with them. The French writer Joseph Gobineau, the father of modern racist ideology, in his 1854 "Essay on the Inequality of Human Races" wrote that "everything great, fruitful and noble in the work of man on this earth springs from the great Aryan family." He persuaded himself that the Aryans had spread to found the cultures of ancient Egypt, Rome, China and even Peru and that "all civilisations derive from the white race." Thor Heyerdahl's famous voyage across the Pacific in search of the founders of the civilizations of Polynesia can be traced directly back to Gobineau's writings. They gave rise to a long series of attempts, Heyerdahl's being just one, to trace historical links between cultures (such as those of the Celts and the Incas) which share sun worship, massive stone monuments, and mummies. All were supposed to trace back to the Aryans, who were often equated with the ancient Egyptians.

Anthropology is the study of the movement of peoples, genes and cultures. These were once all thought to be the same thing. Observing one's fellow citizens in the street makes it obvious even to an anthropologist that everyone does not belong to a single racial type: people look different. Difference usually means classification; and it is only a tiny step from classifying people to judging them. The early evolutionists did not hesitate. Blumenbach, who coined the term "Caucasian," was glad to show where his sympathies lay. Part of his definition of this group was ". . . the most beautiful race of men . . . Nature has lavished upon the women beauties which are not to be seen elsewhere. I consider it impossible to look at them without loving them." The views of his successors were just as predictable—people like them at the top, those from far away at the bottom. Even Rousseau, who claimed to believe in the goodness of humankind, never suggested that the noble savage was black.

Rousseau was following an ancient tradition of deciding who is worth what. Ninety percent of the names given to themselves by primitive peoples mean "men," "the only men" or "the best men"; that is, we are human, others less so. The Sioux Indians of North America seem to be an exception. The literal translation of "Sioux" is snake, or enemy. However, this is the name given to them by an adjacent tribe (and picked up by the first French settlers). The Sioux themselves call their tribe the "Lakota"—the human beings, or people.

The idea that humanity was once divided into a series of biologically distinct pure lineages which differ in quality had a disastrous impact. It was particularly influential in Germany. The tie of the philosophy and policies of the Nazis to anthropology, and the desire to return to a lost time of pure races, is very clear. The Gesellschaft für Rassenhygien (Society for Race Hygiene) was founded in 1905. By 1908 all mixed marriages in German Southwest Africa (now Namibia) were annulled and those involved deprived of their German citizenship. Haeckel himself, the German champion of *The Origin of Species,* wrote that "The morphological differences between two generally recognised species—for example between sheep and goats—are much less important than those between a Hottentot and a man of the Teutonic race." This philosophy ended in the disaster of Hitler's racial policy.

There are ties between the biology and the politics of human differences which began before Hitler and were not broken until many years after his death. Until 1923 the Statue of Liberty really did welcome, as its inscription says, the "huddled masses, struggling to be free." In his 1916 book *The Passing of the Great Race* the euphoniously named American, Madison Grant, echoed many of his fellows when he complained that alien races were being grafted onto the American racial stock. With the advice of biologists, President Coolidge was moved to say that "biological laws tell us that certain divergent peoples will not mix or blend. The Nordics propagate themselves successfully. With other races, the outcome shows deterioration on both sides."

After determined genetical lobbying, the first Immigration Act was passed in 1924. It set limits to ensure that the ethnic composition of the U.S.A. stayed at what it had been in the late nineteenth century.

Each country was allowed a quota of two percent of the numbers of its citizens living in the United States in 1890 (when most immigrants were from the British Isles, Scandinavia and Germany). The law was very good at keeping Eastern Europeans out and left many of them to the mercies of the other experiment in race hygiene which soon began there. It was not repealed until 1966. The theory of pure races had cast a long shadow. Its specter has not yet disappeared. A Hungarian political party, campaigning against rights for gypsies in 1992, described them as a "disadvantaged group, to whom the laws of natural selection have not been applied."

Genetics has at last provided the tools to test the pure race theory. The word "race" is woolly and ill-defined. As it includes social and political as well as biological criteria, genetics alone can never claim to have solved the problem of human racial differentiation. In an attempt to escape a problem by redefining it the term "ethnic group" is sometimes used. Part of the problem is that such groups can define themselves. The Scots scarcely existed until they were invented by King George IV, who in 1822 visited Edinburgh and, dressed in a Stuart kilt and a pair of flesh-colored tights, gave the Scots a national identity they never knew they possessed. It took only the imagination of Sir Walter Scott in devising a native culture to produce a new and potent ethnic myth. Much of it was based on the kilt, which, as Macaulay said, "before the Union, was considered by nine Scotchmen out of ten as the dress of a thief." The Celts—the larger unit to which the Scots are supposed to belong—are themselves largely an illusion based on the incompleteness of the archaeological record.

For ethnic identity what matters most is what group we think we belong to. For genes it is not so simple. It is possible to argue that the genes which matter are those we see—after all, people do tend to choose mates of the same skin color as themselves, so that this is what counts when it comes to discussing race. However, the theory of pure races made a definite claim about human groups: that they descend from a series of distinct ancestors. If this is so, and the genes which change people's appearance really do represent the remnants of this history, then the races of the world should be distinct from one another in a large sample of their genes and not just those for skin color.

This leaves the problem of which genes to use. Information on DNA sequence might be useful. As the majority of the DNA has no productive role and as arguments about race usually degenerate into discussions about biological quality, it makes more sense to look at functional genes, such as blood groups, enzymes and proteins. There is also far more data available on these than on the order of the bases in the DNA. Information on DNA is beginning to emerge and is generating its own controversies about the use of genetics in deciding racial affiliation.

We already know that for proteins or blood groups no two people within any population are the same. What does the genetic atlas look like? Are the trends in skin color—resulting, as they do, from changes in less than ten genes—accompanied by parallel trends in the hundred thousand functional genes which make up a human being?

Everyone can see the global trends in skin color, hair form and so on. There are plenty of less obvious patterns. The reason for most of them is quite unknown. For a few, natural selection may be at work (although it is fatally easy to make up stories about its wonders which can explain—or explain away—any pattern of gene distribution). There are certainly a number of global shifts in skin color, body shape, and hemoglobin structure which do seem to have evolved through the action of selection.

Other patterns are so striking that they almost beg to be explained in the same way. The blood groups are a case in point. In England, the gene for group B in the ABO system is fairly rare—less than one in ten people carry it. In central Russia and West Africa B is common, with up to a third of the population possessing this variant. The pattern might reflect different susceptibility to disease, but this has never been proved. In the rhesus system, marriages between a positive man and a negative woman can be dangerous as the mother's blood may react against that of her child. Nevertheless, rhesus negative is common in Europe and Africa. It must presumably have some advantage which has allowed it to spread in the face of this obvious penalty.

Even the most imaginative would be pressed to come up with a selective explanation for most geographical trends. For example, most westerners have wet and sticky ear wax, but the wax of most orientals is flaky and dry. The new biology is producing other equally baffling differences among the various parts of the globe. Europeans have always been drunkards. Literature is full of references to the pleasures of alcohol. Poison though it is, most of us cope quite well with drink. Our liver enzymes break down the alcohol into a form which is easily disposed of. Alcoholics can be treated with drugs which stop the enzymes from working, so they feel weak and sick after even a small binge and soon learn to avoid drink.

In Japan, most people avoid alcohol. They do so because when they drink, their face goes red and they feel ill. They have a variant of the liver enzyme which is much worse at getting rid of the poison than is the form found in the West. This version is less common in Japanese alcoholics, who tend to carry the western form of the gene. Disulfiram, the drug used to treat western alcoholics, causes symptoms similar to that of Japanese when they indulge in their favorite hobby. The trend in the frequency of the alcohol-metabolizing gene is from west to east. Again, the reason is quite unknown.

As so much of modern medicine—blood transfusion, tissue transplants and the treatment of inborn disease—depends on genetics we have suddenly arrived at the rather surprising position of knowing more about the patterns of genetic change in humans than in any other animal. Hundreds of different genes—for blood groups, enzymes, and inherited variants on the surfaces of cells—have been mapped. Most, like skin color, blood groups or alcohol tolerance, vary from place to place. The picture which emerges is quite different from that supported by those who believe that the human species is divided into distinct races, distinguished by skin color. In fact, the trends in skin color are not accompanied by those in other genes. Instead, the patterns of variation in each system (be it blood group, enzyme, or cell-surface antigen) are largely independent. We would have a very different view of human race if we diagnosed it from blood groups, with an unlikely alliance between the Armenians and the Nigerians, who could jointly despise the B-free people of Australia and Peru. When gene geography is used to look at overall patterns of variation it seems that people from different parts of the world do not differ much on the average. Color does not say much about what lies under the skin.

Imagine that the whole of the world's population is measured for the total amount of genetic diversity that it contains for blood groups, enzymes and cell-surface types. The job should be easy enough; after all, the people of the world would all boil down into a soup which would just fill one of the smaller Finger Lakes in upstate New York. The total set of variation can then be sorted out among people, countries and races to see how it splits up.

The analysis—which is based on eighteen variable genes in a hundred and eighty different populations—shows that around eighty-five percent of total diversity for this sample of genes, worldwide, comes from the differences between different individuals from the same country: two randomly chosen Englishmen, say, or two Nigerians. Another five to ten percent is due to the differences between nations; for example, the people of England and Spain, or those of Nigeria and Kenya. The overall genetic differences between "races" (Africans and Europeans, for example) is not much greater than that between different countries within Europe or within Africa. Individuals—not nations and not races—are the main repository of human variation for functional genes. A race, as defined by skin color, is no more a biological entity than is a nation, whose identity depends only on a brief shared history.

Geographical changes in genes show that the idea that humanity is divided up into a series of distinct groups is wrong. The ancient private homeland in the Caucasus—the cradle of the white race—was just a myth, as were its equivalents in Egypt or Peru. If, after a global disaster, only one group, the Albanians, the Papuans or the Senegalese, survived, most of the world's biological diversity would be preserved. Humans are a rather homogeneous species, perhaps because they evolved so recently. Using information on variation in a sample of proteins as an indicator, the difference among the races of humankind is only about a fiftieth of that between man and chimpanzee. This reflects the short time that has passed since humans began to fill the globe and to diversify into the groups found today.

Other creatures vary much more from place to place. Race really does mean something to them. The genetic differences between the snail populations of two adjacent Pyrenean valleys is much greater than that between Australian aboriginals and Europeans. That between the orangutan of Borneo and that of Sumatra, just a few miles apart, is ten times greater than the difference between any pair of human groups, probably because they have been evolving independently on the two islands for so long. For a snail or an orangutan it makes good biological sense to be a racist, but humans have to accept the fact that they belong to a tediously uniform species.

Relatively uniform we may be, but some global patterns do exist. Many differentiate Africans from the rest of the world's populations. For example, Africans as a whole are more diverse than the other peoples of the world, probably because humanity escaped from Africa rather late in its history. Non-Africans represent just a small sample of the genes in their native continent. The genes also suggest that most non-Africans are more closely related to each other than they are to the populations of Africa.

The fact that genes can be used to differentiate peoples—such as

Africans and Europeans—is scarcely relevant to how different they actually are. After all, a forensic scientist can separate two brothers suspected of a crime on the basis of a blood sample, although the suspects share half their inheritance. Even a single gene (which represents a tiny part of the total difference between two people) may be a reliable indicator. If a bloodstain at the scene of a crime contains sickle-cell hemoglobin, it is almost certain that the suspect has African ancestry; but if there is the gene for cystic fibrosis (which is unknown among Africans) then the police should look for a European. Neither observation changes the fact that Africans and Europeans have, on the average, most of their genes in common.

The issue of differentiability versus difference is causing new controversy. DNA fingerprints are enormously variable. Everyone is unique. When they were first discovered, astonishing claims were made about how they would revolutionize forensic science. In one American court, the prosecution described the chance of being wrong as one in seven hundred and thirty-eight million million. A single trace of DNA—blood, sperm, or even the root of a hair—and the suspect would be identified. There was, it seemed, no room for argument. The case was so persuasive that sometimes judges even refused to hear evidence from the defense which challenged the method.

Now, things look rather murkier. First, of course, there is the simple fact that even if the test is infallible, the people who make it are not. There have been some obvious lapses (such as mistakes in labeling the samples being tested). Other technical problems can also lead to difficulties. When DNA from the scene of a crime is compared with that of a suspect, the stained bands of each sample are lined up and compared by eye. Because the eye is an unreliable instrument there is plenty of room for error in deciding whether or not the bands are identical. There have been courtroom battles between defense and prosecution experts about just what "identical" means in this context. Courts now group similar bands together into "bins" to reduce the chance of mistakes.

These arguments are the stuff of legal dispute and are no different from the controversies about other forensic tests (such as those for explosives) which often hit the headlines. However, there is a more fundamental problem in legal genetics, which arises from evolutionary history and from the differences among the groups of humankind.

DNA fingerprints are made up of short sequences of the DNA message which are repeated again and again. The number of repeats and the position in which they occur varies from person to person. This is what gives the method its specificity. Usually, a sample from the scene of the crime is compared with a sample from the suspect and with some from a "panel" of innocent blood donors. The procedure is rather like a police lineup, in which witnesses pick out the criminal from a group known not to have committed the crime.

In the earliest days of DNA fingerprinting the Federal Bureau of Investigation set up a reference group of innocent DNA donors made up of white police officers. To some jurors, if the suspect's fingerprint was much more similar to that at the scene of the crime than to that of every member of the panel, the case seemed indisputable. The defendant had committed the crime.

In fact, there is a potential problem in this apparently simple approach. If an eyewitness had seen—say—a white person committing a crime, and then had to pick out the alleged criminal from an identity parade consisting entirely of blacks, legal eyebrows would be raised. Obviously, the ethnic group of the suspect has to be matched with that of the group with which he is compared because the clue (skin color in this case) differs genetically among the peoples of the world.

DNA fingerprints have a very high mutation rate and evolve quickly. Since they were discovered it has been found that those from people with African ancestry are somewhat different from those of Europeans (although the overall genetic divergence between Africans and Europeans for this character is no greater than that for enzymes and blood groups, with nine-tenths of total diversity due to differences among individuals within racial groups). For example, there is more variation in the number and position of the repeated DNA sequence used as the basis for fingerprinting in Africans than in Europeans. As a result, African DNA is sliced into a wider range of lengths (some of which are longer than those from any European). The fingerprint pattern of people with African ancestry may hence be noticeably different from that of a typical European. In South American tribal populations there are geographical differences on a smaller scale too.

This raises several potential difficulties. To take an extreme example, imagine a black suspect who is—wrongly—accused of a crime which was in fact committed by another black man. The DNA fingerprint of the suspect is compared to that left at the crime and to those of a panel of white policemen. The genetic divergence between the races means that the innocent suspect's DNA is inevitably more like that of the criminal than that of any European. There is hence a danger that he will be wrongly convicted. Also, if Africans or any other group have their own characteristic set of fingerprints and a DNA band common in that group is found at the scene of the crime the sample is likely to have other bands specific to that population. This confuses the estimation of how likely it is that the suspect's similarity to that of the bloodstain has arisen by chance. There are now three different data bases in the United States from Caucasians, Hispanics and African-Americans in an attempt to get around the problem.

All this has led to controversy in the world of DNA fingerprinting and it is right that it should. In the United States at least, where capital punishment is common, the issue may be one of life and death. The rule in American courts is that scientific evidence may be rejected if it is not generally accepted in the scientific community. Several research papers have appeared casting doubt on the assumptions used in calculating the chances of a mistaken match. Appeals courts in California and Massachusetts have thrown out convictions for murder and rape because they are not satisfied that DNA fingerprinting is "generally accepted" by scientists. A British court did the same thing in a case of armed robbery when the defense produced an American statistical expert. There is now a sprint to gather information on DNA fingerprints from all over the world so that at least a suspect can be compared with someone from his own local group and the chances of false matches worked out properly. It does seem that any differences among races are too small to lead to real concern that the method is untrustworthy; but it is equally clear that the enormous claims made about its power in its early days were based on ignoring the genetic differences which do exist among the globe's populations.

Although people from different parts of the world differ from each other, the idea of pure races is a myth. Much of the story of the genetics of race—a field promoted by some of the most eminent scientists of their day—turns out to have been prejudice dressed up as science, a classic example of the way that biology should not be used to help us understand ourselves. The moral issues raised by our own biology—racism, sexual stereotypes, and claims that selfishness, spite and nationalism are driven by genes—are just that: issues of morality rather than science. Science has, or should have, nothing to do with how we perceive or treat our fellow human beings. Although it is comforting to the liberal conscience that genetics (at least the little we yet know about the subject) does show that there are few real differences among the peoples of the world, this is irrelevant to the issue of racism, which is a moral and political one.

This means, of course, that those who are determined to dislike one race or another are unlikely to be much impressed by scientific arguments. I once gave a lecture on race when I was teaching African undergraduates in Botswana. The class was delighted to learn that they were not much different from the white South Africans who despised them so heartily. At the end of the lecture there was just one question. Surely, a student asked, what you are saying can't be true of the Bushmen; they are *obviously* different from the rest of us. I must admit to a certain feeling of despair at that.

My own view is that although biology may tell us a lot about where we come from, it says nothing about what we are. The dismal history of racial genetics strengthens this belief.

14 | Evolution Engineered

MOST BIOLOGISTS have not read *The Origin of Species*. The same is true, no doubt, for most Marxists when it comes to *Das Kapital*. After many years of studying evolution and constantly referring to Darwin's ideas I first perused *The Origin* on a Greek beach to ward off the unbearable tedium of being on holiday. It was a remarkably easy read. However, the first couple of chapters were a surprise: far from being a deep account of the philosophy of existence, or even the theory of evolution, they are mainly about pigeons.

Darwin went to great lengths to show how animal breeders had, by selecting the birds they liked best, produced varieties as different as the roller, the tumbler and the pouter from that rather ordinary bird the rock dove. Exactly the same process had created breeds of domestic cattle, of dogs and of horses. Darwin used the results of those who had applied evolutionary ideas without realizing it to show that his theory actually worked. He went further than the pigeon fanciers only in suggesting that, if selection went on long enough, barriers to genetic exchange would arise. A new form of life, a new species, would be born.

Evolution is now an applied subject in its own right although many of those who use it still do not realize quite what they are doing. Until recently the methods of applied evolution, or biological engineering, have always been close to those of life itself. In life, tinkering works, and, given enough time, can be the means to an unexpected end. All technical advances once had this utilitarian approach. The engineers who designed stone tools or steam engines had no idea of the physics of how their machines worked and the first farmers developed new crops with no knowledge of heredity at all. Pragmatism led to progress, much as it had throughout history.

Nowadays, engineers have a very different world view. Their philosophy is to plan ahead and design what is needed, using as much scientific theory as is necessary. Applied biology, from agriculture to medicine, has adopted this approach only in the last few years. By so doing it is at the beginning of an advance as spectacular as that of transport since the steam engine.

The fusion of Mendelism and Darwinism has already made farming much more productive. The amount of food available per head, worldwide, has gone up in the face of the greatest population explosion in human history. This success has already brought problems and, if one thing is certain about the new attempts to engineer nature, it is that nature will respond in unexpected and unwelcome ways.

Mendel or Darwin would feel quite at home with some of the new technology. The "Green Revolution" is based on traditional methods of plant breeding. One of its most effective tools is to use stocks of rice and wheat with stiffer and shorter stalks than normal. Just a few genes are involved. Dwarf varieties were crossed with others with particularly rigid stems. Their descendants were mated with stocks containing genes for high yield and rapid growth. Plants which had the best qualities of the parental types were chosen and the process continued for several generations. To use Thomas Hunt Morgan's term (see p. 57), these plants were recombinants; they contained mixtures of characters (short stem and high productivity) never previously found in nature. At a stroke, one of the main problems of tropical agriculture, the tendency for rice and wheat to grow tall when fertilizer is used but to fall over in high winds, was solved.

This simple trick transformed the rural economies of India and China. Directed evolution in less than fifty years gave a sixfold boost in crop yield. The increase in production was as great as that at the origin of farming ten thousand years ago.

Just the same approach works in animals. *The Origin of Species* itself describes the improvement of sheep using a "short-stemmed" mutation. The Ancon gene shortened the legs of the sheep who bore it. This was convenient as it stopped them from jumping over stone walls, and the breed became popular. Now, alas, with the invention of barbed-wire fences it has disappeared. Other useful genes—for disease resistance in tropical cattle, increased growth in pigs, and so on—have been bred into farm animals and spread by selection in the Darwinian way. Often improvements depend not on choosing single genes, but simply on breeding from the best (which usually involves changes at many genes at once). The results can be spectacular. After all, the poodle and the St. Bernard had a common ancestor only a few thousand years ago. In 1904 an experiment was started in Illinois in which, each generation, maize plants particularly rich in oil were bred from. The experiment still goes on and, nearly a hundred generations later, the average amount of oil per plant has gone up by a dozen times, with no sign of any slowing of progress. Straightforward applied evolution (which involves nothing more radical than changing the direction of natural selection or bringing new mixtures of genes together) can do remarkable things.

Another way of refining Darwinism is to increase the flow of genetic raw material upon which it feeds. More mutations may mean quicker progress. Penicillin production once depended on tiny amounts of antibiotic from large vats of fungus. By breeding from the most productive strains there was a hundredfold increase in yield. The next step was to mutate the genes involved, using radiation and chemicals. A whole new generation of antibiotic drugs soon emerged. The same approach has had great success in improving tomatoes.

The supply of genes can be boosted in another way. The wild plants from which today's crops were derived are full of potentially useful variation. In modern farming, as in modern life, efficiency is gained at a price. Most crops are highly inbred. Each plant has exactly the same set of genes. They have reached an evolutionary dead end as no variability is left. However, their enemies, climate and disease, are not so constrained. During the 1960s there was a series of expensive disasters in the grain combines of North America. Nearly all of them had turned to growing the same variety of corn. Suddenly, fungi evolved to overcome its resistance to disease and millions of acres were wiped out. In 1970 a sixth of the whole crop, worth billions of dollars, was destroyed. This sparked off a frantic search for their semi-domesticated ancestors (which retained many of the genes for disease resistance which had been bred out from the modern crops). Expeditions were sent to the Middle East and to the other great centers of plant diversity such as the Andes to find native stocks before they were replaced by western varieties. Although many genes had been lost forever, there are now seed banks (some in unexpected places such as in the cold dry air of

Spitsbergen) for most crops. They contain a mass of inherited variation, the fuel upon which biological engineering depends. Who owns this precious resource is another issue: at present, as in the exploitation of Africa by the nineteenth-century colonial powers, genes are exported without much benefit to local people. Plenty of grains are preserved in excavations of the farms and homes of peoples long dead: the DNA which codes for a protein improving bread quality has been extracted from some wheat seeds collected in a British Iron Age fort two thousand years old. It is still a long way before such genes could be re-inserted into modern crops, but perhaps some of the inherited diversity driven to extinction by the spread of modern varieties has been spared in an unexpected way.

The standard agricultural approach of breeding from the best—evolution writ large—has limits, which are often reached. The North American maize crop and some of the more disgusting breeds of domestic dog can evolve no further: they have used up their genetic reserves.

The most important constraint on evolution is set by sex: the fact that to make creatures with new mixtures of genes their parents must mate. There are strict biological controls as to who can mate with whom. The partners must, of course, be of different sexes, but they must also come from the same species. This is the best way of defining what a species actually is: two individuals belong to the same one if they can combine genes in their offspring. To recombine genes in nature or on the farm sex is unavoidable.

Although this restriction seems inevitable it decreases the raw material available to the evolutionary engineer. Genes which might be useful in improving one creature cannot be used because they are locked away within another. Usually there is no way of getting through the reproductive blockade. Species put limits on genetic free trade. They mark biological frontiers which mean that a profitable gene which emerges in one species cannot be exported to another.

The biggest advances in applied genetics come from breaking the sex barrier. This is how agriculture itself began. Early farmers ameliorated nature by irrigation or by clearing trees to allow vegetation to flourish. This disturbed the local ecology. In such habitats plants which never normally meet came together. The barriers between plant species are more permeable than those which divide animals and, occasionally, hybrids appeared. They contained combinations of genes which had not been seen before. The process can be seen at work today. Many mudflats around Britain are being covered by a tough grass. It is a hybrid between a local species and one accidentally introduced from America. The new species, a mixture of the genes of its predecessors, is better at coping with a harsh and salty environment than was either of its parents, and is rapidly becoming a pest.

Crop plants, like those who cultivate them, retain their genetic history within themselves. Chromosomes show that modern wheat began when two species of grass hybridized. Each survives today in the Middle East and produces seeds which can be used for food. As in the estuarine grass, the hybrid was more productive than either parent. Perhaps the barrier to mating was broken by chance; or one species may have been planted among natural populations of the other. However the hybrid was formed, the farmers of ten thousand years ago quickly made use of it. Soon, another grass crossed with the new crop, improving it further. This was the predecessor of every one of the billions of wheat plants grown today. The new crop contained a wider range of genes than any of its ancestors. Inadvertently, the early farmers had moved chromosomes, genes and DNA from one species to another. They were the first genetic engineers.

Now, trading genes between species in this way no longer depends on good luck. Planned parenthood is involved. The new crop triticale is a hybrid between wheat and rye. It can grow in hot dry places and will be of great benefit to tropical agriculture.

Triticale and the other hybrids are just the first step toward a free market in genes. Genetic engineering is a way of circumventing sex altogether. Molecular biology makes it possible to shift genes among lineages which are normally insulated from each other; to make recombinant DNA without bothering with sex. Improvements in technology mean that genes can be moved from—more or less—anywhere to anywhere else. Already, they are routinely transferred between creatures as distinct as humans and bacteria. DNA can be used throughout the living world, wherever it comes from.

Genetic engineering began in bacteria, which have a commendably wide range of sexual interests. They exchange genes in many ways; by taking up naked DNA, by a process of mating rather like that of higher animals and by using a whole range of third parties, such as viruses, to transfer DNA. This "infectious heredity" (which suggests that venereal disease evolved before sex) has been subverted by science.

The gene to be engineered (which may be from another bacterium, from a plant or from a human being) is inserted into a piece of viral

DNA using various technical tricks. The manipulated virus plus its new gene is used to infect a new host. With luck, the recipient will treat the immigrant DNA as its own, making a copy every time its cells divide. In this way bacteria can be persuaded to make vast numbers of copies of the engineered gene—and vast amounts of whatever it manufactures: pure human proteins, drugs, or an array of other things. The same methods can be used on plants, animals and even humans. A new field of applied Darwinism has been born.

Crossing the sexual divide, deep though it is, between bacteria and other creatures proved surprisingly easy. One of the first human genes to be used was that for the hormone insulin. Insulin was once extracted from the pancreas of pigs. Now the human gene has been cloned and large quantities of the pure protein can be produced. Human growth hormone, too—once extracted laboriously and with much controversy from the pituitary glands of the recently dead—is now made in the same way. This gets around a macabre and unexpected problem. A few patients caught a nervous degenerative disease from corpses carrying a virus. The story of this medical disaster is close to that of those infected with AIDS when factor VIII from donated blood was used to treat hemophiliacs. Now, the factor VIII gene, too, has been inserted into bacteria and some patients are treated with the bacterial product.

Genetic engineering can be used against infectious disease. Jenner could use the cowpox virus to vaccinate against smallpox (an experiment which would fall foul of even the most lenient Ethics Committee today) because the two viruses share antigens, cues of identity recognized by the immune system as the basis of its defensive response. Antibodies against cowpox protect against smallpox. However, there are dangers; cowpox itself can cause problems and even in modern vaccines there is a small risk of infection or of a reaction to the injection of foreign proteins. Many diseases (such as leprosy) cannot be helped by vaccination because it is hard to grow their causative agents in the laboratory.

A cunning piece of engineering gets around the problem. The antigen genes from an agent of disease are inserted into a harmless bacterium. There is no risk of accidentally spreading the disease as the genes for virulence have been left out. Antigens from several different sources can be inserted into the same bacterium, to give a single vaccine against many infections at once. A modified strain of Salmonella (which in its native state can cause food poisoning) is used. The bacterium, with its added antigens, flourishes for a short time in the gut. By persuading the host that it has been infected it ensures that antibodies are made. A vaccine made in this way is being used to treat wild foxes in the hope of slowing the spread of rabies across Europe.

Engineering a way through the sex barrier is also important in agriculture. Huge sums of money can be made by increasing crop yield and huge amounts of work are being done by those who hope to benefit. Some of the tricks are simple. Plants can make copies of themselves from only a few cells. This makes it possible to produce many plants from one without bothering with sex at all. It is hard to improve trees by breeding from the best, because it takes so long. By finding a superior specimen and breaking its tissue into single cells, copies of that super-tree can be grown to give, within a single generation, a super-forest. The process is used to grow improved oil palms and there is hope of replacing the elms which once filled the English countryside (and were devastated by Dutch Elm Disease) with clones resistant to infection. A related method may replace natural vanilla, now extracted at great expense from a tropical orchid, with the same chemical extracted from cultures of vanilla cells grown in the botanical equivalent of a factory farm.

Genes can also be inserted into plants. As most lack certain amino acids it is hard to stay healthy on a strict vegetarian diet. Much could be done by moving the right genes in. Many plants produce powerful natural pesticides—not surprisingly, as they are at constant risk of attack. Some of them, coffee, cocaine and pepper for example, are used as drugs for pleasure or profit. Now the pesticide genes from one species can be shifted into another; which may cut down the use of toxic sprays. Another trick is to introduce a gene which makes the plant resistant to weed killers. The field can be sprayed, killing the weeds but leaving the crop untouched. Plants can even be "vaccinated" by introducing a few of the genes from their natural virus enemies. When the virus strikes, it uses the plant's machinery to make copies of itself. If parts of its own structure are already being made, this interferes with the copying mechanism and the attack fails.

We grow plants because they make useful products—food, for example. They could be used as much more versatile biological factories. There is already the prospect of using potatoes and tobacco plants to make antibodies and other human blood proteins.

The glittering prize for the agricultural engineers is to introduce genes which allow crops to make their own fertilizers. Clover has already evolved an arrangement with certain bacteria. The bacteria take nitrogen from the air and turn it into a form which can be used by the plant. In return they gain food and protection. Farmers plant mixtures of grass and clover which are more productive than either grown alone. Putting nitrogen-fixing genes directly into crops would dramatically reduce the need for fertilizers. No one has yet succeeded in making the right bacterial genes work in a plant cell. The rewards for doing so are huge; and no doubt there will be success one day.

All this may mean that plants will soon do almost everything and that animals will fade in importance as—perhaps—the salmon-flavored banana takes over. No doubt a few unregenerate carnivores will remain. Applied evolution can help them, too. Cow embryos are made in the laboratory by fertilizing desirable eggs with superior sperm, allowing them to divide and splitting them into smaller portions which are introduced into new mothers (who need not themselves have any particular merit). This multiplies the number of high-quality calves. It is easy to freeze the embryos until they are needed and surrogate motherhood is already used on thousands of cows each year. Perhaps it will become possible to use adult cells in the same way. The rural landscape may become one in which asexual cows feed on engineered grass under the shade of clonal trees.

Foreign DNA can be introduced into animal cells, too (although the process is not as easy as in bacteria or plants). Body cells or eggs can be used: in the latter case the gene may be passed on to later generations. Already, genes for human proteins—such as one of the hemophilia clotting factors—has been introduced into sheep so that the pure protein is produced in their milk (giving a new rural pastime called "pharmaceutical farming"). Mice with human growth hormone genes inserted grow up larger than normal. The same DNA has been put into pigs, but although the animals grow quickly they are unhealthy. Fish are easier to handle. Their large eggs take up foreign DNA and the growth hormone gene can make hatchery fish grow more quickly. Another ingenious idea is to introduce the DNA which codes for a natural "antifreeze" found in arctic fish into tropical species, enabling them to grow in northern waters. This gene has even been inserted into tomatoes helping them to resist frost. Insects, too, can be engineered. It will soon be possible to insert genes for insecticide resistance into useful creatures (such as a mite which attacks crop pests) allowing crops to be sprayed without destroying the pests' natural enemies.

All this is all very well. However, to interfere with the boundaries between species can stir a deep unease. It has met with resistance, up to and including some fairly decorous German riots. Part of the problem is the word "engineering," which contains more of a threat than does the cozy term "domestication" used of the first genetic engineers. Part comes from the caution of biologists themselves who, at the beginning of the new age twenty years ago, froze new experiments until safety rules were worked out. There is also a fear of germs, based on the idea that all bacteria are harmful. They are in fact essential, each one of us containing ten times as many bacteria as we have cells of our own. Most important, people are suspicious of technical fixes, the idea that technology can overcome all problems. From nuclear power to irrigating the desert the optimism of engineers has often turned out to be a short-lived thing.

There are concerns about economic side effects, too. The Green Revolution, although it improved food production, drove farmers from the land as large companies gained control of seed production and the sale of fertilizers. Much the same happened in the early days of American farming. In the 1930s there appeared new strains of hybrid corn which—as at the beginning of agriculture—were made by crossing two lineages together. Their sale was controlled by a few combines who by manipulating the price drove many small farmers out of business. Genetically engineered stocks (which are protected by patent law) pose the same danger of a new harvest of the grapes of economic wrath. Few farmers could bargain with an organization which had a monopoly on the sale of a herbicide-tolerant plant—and sold it in consort with the herbicide involved. It also makes little sense to spend money increasing the numbers of dairy cattle by embryo transfer when there is already a surplus of butter, or genetically manipulating wheat to add to the grain mountain.

The most widespread fear is of the escape of genetically manipulated creatures who may unleash a new plague upon the world. Biologists have some standard defenses to this concern. Genetically manipulated creatures are likely to be less fit than those which have not been interfered with. After all, if the gene gives its carriers an advantage it might be expected to have evolved by natural means. The unfitness of artificial creatures is already obvious. Most farm animals and plants cannot survive outside farms, which is why the streets are not full of marauding cows, sheep or potatoes. The same is likely to be true of bacteria and viruses. In Britain and the United States, children are injected with live polio virus which has been "attenuated" to make it less dangerous. Surveys of sewage show that this live virus is constantly escaping. This is the key to the scheme's success: even children whose parents do not allow them to be vaccinated are exposed to the virus emerging from their friends who have just been treated. However, the attenuated virus has never survived in the wild; it depends on a constant supply of newly vaccinated children. If all genetically engineered organisms are as feeble as the polio virus, there is not much to worry about.

Nevertheless, it is worth remembering that almost every domestic animal is a pest somewhere. Cats wiped out most of New Zealand's birds. Goats have done the same or worse in many places, feral pigs are everywhere in the subtropics and even horses can be a nuisance as they roam the California deserts. Plants are even more destructive. Everyone knows about the prickly pear in Australia, but a pretty yellow South African garden plant, the soursop, is doing enormous damage to grazing lands. Wherever domesticated creatures have escaped, native plants and animals have suffered.

The brash biologist can—and does—argue that we know enough not to repeat these early mistakes. Biologists also point out, quite correctly, that much of what genetic engineering does is perfectly natural. Recombinant DNA is made every time sperm meets egg; species are not fixed entities as they evolve from one into another and—regularly in bacteria and sometimes in plants—they even exchange genes by natural means. Huge numbers of bacteria are constantly produced, the human race alone excreting a total of ten with twenty-two zeros after it each day. Because of mutation, many must consist of genetically new forms and a few, through the vagaries of their reproduction, must include genes incorporated from other species. None has spread and gut bacteria remain benign.

Viruses, though, are slightly less comforting. Some of the flu epidemics which pass through the world each winter originate in China (p. 171). The human flu virus sometimes incorporates other genes which come from the viruses of wild birds. Only by passing from farm ducks to pigs to humans do the new mixtures of genes cause trouble, but they are a salutary reminder of our own vulnerability to rare events in distant places.

However, those in charge have now been persuaded to permit the release of a few genetically manipulated organisms. In California, crops are damaged by frost. As the air cools, small patches of ice appear on the leaves around natural colonies of *Pseudomonas* bacteria. One bacterial gene is responsible for this irritating behavior. Occasionally it changes by mutation to produce an "ice-minus" strain which does less harm. Now an artificial ice-minus bacterium has been made, which, when sprayed onto plants, cuts down frost injury by displacing the icy natives. The gene was removed from a normal bacterium, sections cut out and the altered DNA reintroduced into a *Pseudomonas* stock. Although the bacteria are in some senses not engineered at all as the genes involved come from their own species, the plan caused an uproar and was delayed for several years. This irritated agricultural researchers. As they pointed out, legal controls would prevent moving DNA from a weed to a crop to improve it—which is what happened at the beginning of farming when the first wheat was made. After many court-room battles the release was allowed (largely because ice-minus bacteria have turned up thousands of times by natural mutation with no apparent harm).

During the court cases about ice-minus it emerged that the military had already done dreadful things without the public being allowed to know. Biological warfare was once a popular excuse for spending more on defense. What the army really wanted to study was how best to infect people. In the early 1950s huge numbers of *Serratia marcescens* bacteria, then thought to be harmless, were sprayed over San Francisco and other American cities to investigate how they spread. Since the experiment it has been learned that *Serratia* can infect those already debilitated by disease and a number of—then mysterious—infections at the time were due to the bacterium (although they have never been tracked down to the strains sprayed by the army). The experiment shows that even a perfectly natural bacterium which appears to have no ill effects may be dangerous when placed in unnatural circumstances.

There are other dangers in genetic engineering. What if the new gene gets out of its own species and into another? Herbicide resistance genes might get from crop plants into their weedy relatives (which, for crops like oil-seed rape, occasionally hybridize with them), giving a new super-weed resistant to spraying. Fish farming involves so many escapes that the genetic structure of North Atlantic salmon has already been damaged by breeding between farmed and local populations. What might happen if the anti-freeze gene allowed escaping tropical fish to displace or hybridize with the natives?

Although some fears are exaggerated, to release genetically manipulated creatures is to play with the unknown—and hence, inevitably, to take a risk. Some scientists' defenses suggest that the risk is so tiny as to be not worth considering. They are still in a phase of technological absolutism. Trust us, they say, and nothing will go wrong. They sound uncannily similar to the engineers who developed nuclear power, or drained the Florida Everglades. Like the Bourbons, certain biologists seem to have forgotten nothing of the successes and learned nothing from the disasters of previous occasions when a science evolves into a technology.

A few enthusiasts even disregard the nature of their own subject. They claim that inadvertently creating a monster by genetic engineering is no more likely than making a television set by randomly mixing electronic components. In this they echo a standard creationist argument which is (to quote Duane Gish, one of the founders of the Institute for Creation Science) that the chances of an organ as complex as an eye arising without divine intervention are the same as those of a whirlwind building an airplane as it blows through a factory.

Evolution is all about assembling the improbable by tiny steps; and not until the unlikely has been reached do we notice just what it can do. Genetically engineered organisms will, like any other creature, evolve to deal with their new condition. It is fairly certain that some of them will cause problems. Low risk is not no risk. The question is one which is universal in economics—will the benefits outweigh the costs? For genetically manipulated organisms nobody knows as the experiment has not yet been done. There may be a precedent in another much-vaunted piece of biological engineering, the chemical control of pests.

The first modern insecticide, DDT, was introduced at the end of the Second World War to control lice. It was a spectacular success. The optimists were soon in charge. Their approach was that of an engineer: with money and technology one can do anything. Predictably enough, the animals responded by evolving to subvert the technical fix. Nowhere is the danger of certainty seen more clearly than in the fight against malaria, where biological bumbling triumphed over engineering elegance.

After the conquest of lice, DDT was sprayed onto malarial mosquitoes. Victory soon seemed imminent. The number of infections fell dramatically, in Ceylon from millions to scores. The rot soon set in as genes for resistance to insecticides spread through the mosquito population. The counterattack has been so effective that malaria is raging at unprecedented levels. The World Health Organization admits that "the history of antimalaria campaigns is a record of exaggerated expectations followed sooner or later by disappointment."

The parasites, too, have subverted human attempts to engineer them out of existence. Although resistance took longer to develop than in the mosquitoes, in many places the old malaria treatments are now useless as the parasites have evolved means of coping with them.

The standard Darwinian mechanisms of mutation and natural selection have helped insect and parasite to survive. Among the insects, there is a bewildering range of new mutations. Some break down insecticides or stop them from getting in. Some allow the insect to store the poison, some change the shape of the target molecule and others enable the insect to avoid places which have been sprayed.

The parasites, too, have evolved a variety of tactics against their chemical enemies. The antimalarial drug chloroquine was developed in the 1940s. Thirty years ago it worked almost everywhere. In the 1960s resistance appeared in southeast Asia and South America. It is now all over the tropical world. One of the most effective defenses resembles the mechanism used by cancer cells to combat anticancer drugs. A massive amount of a protein involved in transporting material across membranes is produced and this pumps the drugs out of the cell fifty times more quickly than normal. More recently, genes which give resistance to other drugs—sometimes several at the same time—have turned up. The Walter Reed Army Institute screened more than a quarter of a million compounds in the hope of finding a new antimalarial drug. Only two proved suitable. One was mefloquine, and in Thailand eighty percent of the malaria parasites are now resistant to it. In 1991 there were claims that malariologists are now down to the last drug, with nothing new in sight. As a result doctors are returning to quinine and to an extract of wormwood (first used in China a thousand years ago) although these treatments are toxic and not particularly effective.

The history of genetic engineering may, when it is finally written, turn out not to be too different from that of the war against the insects, in which evolution prevailed after some initial setbacks. Insecticides have worked well and continue to do so. Without them, there would have been no Green Revolution and most tropical crops would be uneconomic. Lice might still be carrying typhus through the poorer parts of Europe and malaria killing even more people than it does today. However, the triumph of human ingenuity has not been unalloyed: because living organisms can deal with new challenges by evolving to cope, genetic engineers, unlike those who build bridges, must face the prospect that their new toys will fight back.

15 | Fear of Frankenstein

IN PARAGUAY there is an isolated village with an unusual name: Nueva Germania, New Germany. Its inhabitants look quite different from their neighbors. Many have blond hair and blue eyes. Their names are not Spanish, but are more likely to be Schutte or Neumann. These people are the descendants of an experiment, an experiment in improving humanity. Their ancestors were chosen from the people of Saxony in 1886 by Elisabeth Nietzsche—sister of the philosopher, who himself uttered the immortal phrase "What in the world has caused more damage than the follies of the compassionate?"—as particularly splendid specimens, selected for the purity of their blood. The idea was suggested by Wagner (who planned to visit, although he never did). They were expected to found a community so favored in its genetic endowment that it would be the seed of a new race of supermen.

Elisabeth Nietzsche died in 1935. Hitler himself wept at her funeral. Today the people of Nueva Germania are poor, inbred and diseased. Their Utopia has failed.

The idea of improving humanity by meddling with genes was once popular. It came from Francis Galton. He was the first proponent of selective breeding based on science. As he said: "What Nature does blindly and ruthlessly, man may do providently, quickly and kindly." His American equivalent Charles B. Davenport was blunter. He felt that "Society must protect itself; as it claims the right to deprive the murderer of his life so also may it annihilate the hideous serpent of hopelessly vicious protoplasm." The eugenicists' main concern was to control human evolution. Sterilization was an easy way of reversing what they saw to be undesirable evolutionary trends. Their only means for making evolution move in a direction they approved of was exhortation, which did not seem to work particularly well.

Ironically enough, both Galton's Laboratory for National Eugenics and Davenport's Eugenics Records Office are now world centers for the new human genetics. Together, they—and the hundreds of laboratories which descend from them—are beginning to produce the technology for directing evolution which Galton and Davenport lacked. Already there are answers to many of the scientific, if not the moral, questions which obsessed the eugenicists. This chapter is about the relationship between people and genes now that we have the tools to carry out some kind of eugenical program if we wished.

Genetics has undergone an enormous shift in attitude since its early days. Those involved in it scarcely involve themselves with what their work implies for the future of humanity. They feel responsible to people rather than to populations; to individuals rather than to posterity. The shift has gone so far that biologists are more cautious about how their work should be used than is the public. In a recent poll three out of four Americans found the idea of inserting genes into human sperm or egg quite acceptable but almost no scientist was willing to contemplate the idea. This is a remarkable (and healthy) reversal of roles since the hubris of the subject's early days.

No serious scientist now has the slightest interest in producing a genetically planned society. But the explosion in knowledge means that society will soon, like it or not, be faced with ethical problems of the kind comprehensively ignored by the founders of eugenics. Once more there will have to be discussions about whether we can, or should, make choices based on genes; that is, whether we should make conscious decisions about human evolution. There is already concern about the balance between the rights of individuals and of society, but is there any need to worry about those of future generations? Plato felt that there is a moral duty to the future in that "mankind gains its share of immortality by having children" but another intellectual hero, Sam Goldwyn, dismissed it by asking "what did posterity ever do for me?"

The biggest problem of modern genetics is one which it has scarcely come to terms with: the ubiquity of inborn disease. In some places the problem is already obvious. Around the Mediterranean and in Africa there are many inherited errors in red blood cells, which evolved to protect against malaria. In Cyprus and elsewhere the most common damaged gene is for a form of thalassemia, the loss of a segment of the hemoglobin molecule. A single copy is useful as it protects against infection but a child born with two has severe anemia. The treatment is blood transfusion, which works but is expensive. In Cyprus to treat all affected children will soak up half the total health budget within ten years. There are already two hundred and fifty million people who carry a single copy of one of the malaria resistance genes and by the end of the century one person in fifteen, worldwide, will be a carrier. Barring a medical breakthrough no society will be able to afford to treat the millions of anemic children who will be born. High cost will mean hard choices.

Even in places without a history of malarial infection there is plenty of inherited disease. In the United States, about one child in thirty is born with a genetic problem of some kind. Over a third of registered blind people are blind for genetic reasons and more than half of all cases of severe mental handicap have an inherited cause. If the definition is extended, as it should be, to include diseases such as cancer or heart disease which have an inherited component, two thirds of the population will suffer from, and possibly die of, a genetic illness.

Attitudes toward such problems vary greatly from place to place. In Ghana, children are sometimes born with an extra finger or toe. Some tribal groups take no notice, others rejoice as it means that the child will become rich; but others, just a few miles away, regard such children with horror and they are drowned at birth. Even Christianity has seen the genetically unfortunate as less than human. Martin Luther himself declared that Siamese twins were monsters without a soul. Decisions based on appraisals of inborn quality are not new.

Many of the choices which must now be faced are simple and are not very different from those posed by Galton. Should all copies of a particular gene be allowed to pass to the next generation, or should the human race attempt to enhance its biological quality in some way? There are each year about ninety million births and sixty million induced abortions. Many more pregnancies end without the woman knowing, often because the fetus has a genetic defect. Genetic selection is a natural part of reproduction. Changing the balance between uncompleted and completed pregnancies has led to bitter controversy. Some demand that the state be allowed to control reproductive choices and make legal abortion difficult, but others feel that such decisions must be the parents' alone.

What judgment to make is changed by people's experience of inborn disease. Sardinia is a rather traditional Catholic society in which many children are born with thalassemia. Nine tenths of couples at risk of having an affected child now know this; and when the woman becomes pregnant nine tenths of them choose to end pregnancies which would produce a genetically damaged infant. In the United States the search for carriers of Tay-Sachs disease (an inherited degeneration of the nervous system most common in people of Jewish ancestry; see p. 180) has met with the same level of acceptance. Testing of older mothers in Denmark has led to a fivefold decrease in the number of Down's Syndrome children born. At least in places where genetic testing is widely available there is the prospect that illnesses such as Huntington's Disease will soon become rare as those at risk decide not to have children. Some of the most enthusiastic campaigners for tests for inborn disease are parents who have had an affected child and have devoted their lives to caring for it. This in itself says something about how genetical decisions are made and where the ethical balance lies.

Most genetic technology remains depressingly simple. It is to identify a damaged gene and offer the choice of a therapeutic abortion to parents whose fetus is found to be affected. All the common single-gene defects can now be detected in this way. There are even prenatal tests for certain inherited susceptibilities to cancer. Much of the agenda of the Human Genome Project must inevitably be to widen the range of conditions for which such choices can be made. One estimate is that, within fifteen years, there will be tests for around a thousand inherited diseases.

This will certainly lead to new controversies. Where should the line be drawn when assessing biological quality? There are reports that in Russia pregnancies have been terminated because the fetus is carrying genes which predispose to diabetes. But diabetes is a disease which can sometimes be treated with insulin. And what about diseases for which no treatment is yet available, but which might be curable by the time the child is in danger of dying? In muscular dystrophy, for example, we are beginning to understand what has gone wrong in the cell machinery and it is not impossible that some treatment may be available within the next couple of decades. As boys born now with the disease are likely to live that long this poses a moral dilemma of its own.

There can also be a subtle tyranny in prenatal screening: the dictatorship of the normal, the pressure to produce an average child. In the

United States this has already led to demands for growth hormone to be given to children with a slight inherited deficiency who grow up a few inches shorter than average and would once have been accepted as perfectly ordinary. Nearly half the United States population finds the idea of "genetic enhancement" acceptable although geneticists themselves are universally opposed.

There are more immediate problems in giving genetic advice. Some of the first attempts to apply the new knowledge ran into difficulties because they ignored social realities. A search for carriers of the sickle-cell gene in the U.S.A. twenty years ago led to great bitterness in the black community. Although carriers are quite healthy except under conditions of extreme oxygen shortage (which most people never experience), some states made screening compulsory. Those carrying one copy of the mutation were discriminated against in jobs and for insurance. Blacks who did not carry sickle cell thought of those who did as less healthy and less happy than did the carriers themselves. Those who inherited a copy of the gene were discriminated against when it came to marriage and, worst of all, there were hints of eugenic attempts to improve the population rather than the health of individuals. In retrospect, the sickle-cell program—although it was conceived with the best of motives—was a model of the way in which genetic information should not be used.

Other schemes also ran into problems. In Sweden in the 1970s, nearly all newborn children were screened to see if they carried a mutation which, among other effects, made them more susceptible to air pollution caused by cigarettes. The motives were straightforward: to prevent their parents from smoking. Its results were discouraging. Even in well-educated Sweden, many parents found to have children with the gene felt that their offspring were ill although tests showed that most of them were normal. Worst of all, the parents smoked more rather than less, probably because of worry. All this led to the plan to screen every child being abandoned.

However, genetic testing is here to stay. It is already becoming routine. Most pregnant women in their late thirties are offered the chance of a test for the chromosomal abnormality which leads to Down's Syndrome (whose incidence goes up quickly with parental age). The latest version needs only a sample of the mother's blood, which contains enough information about the fetus to allow the physician to be almost certain of its chromosomal constitution.

The pervasiveness of genetic imperfection means that there are limits to how far screening can go. For a recessive disorder (one which requires two copies of the damaged gene to show its effects) there are always far more carriers of a single copy of the inherited variant than of two. If a disorder of this kind affects one birth in ten thousand, about one normal person in fifty carries the gene—which means that there are around a hundred times as many copies in healthy people as in those who suffer from the illness. This means, of course, that the idea that one can improve the long-term health of the population by preventing those with inherited disease from reproducing is futile. More important, it shows that almost everyone carries one or more different genes for recessive inherited disease—they each have at least one genetical skeleton in the cupboard. Any mass screening program would provide information which was unwelcome and might not even be much use.

Take cystic fibrosis. In the United States, one child in two to three thousand is born with the condition. This means that about ten million Americans, one in twenty-five, carry a single copy, so that in about a tenth of married couples one of the partners has the gene. In some communities it is even more frequent. Among the Amish in Ohio, one child in six hundred has the disease and one adult in ten carries the gene. For the Amish, the founder effect, with one ancestor carrying cystic fibrosis, explains its abundance. Why it should be so common in other places is less certain. It might be that carriers are more resistant to infectious disease (perhaps tuberculosis, cholera or plague).

Any screening program would turn up many couples in which one partner is a carrier and a smaller—but still substantial—number in which both carry the gene. There is now a cheap and easy test which identifies most carriers. A trial on a thousand people attending a family planning clinic in southern England found several. The usual response was one of surprise rather than worry. However, if thousands of married couples were routinely screened for all testable recessive genes so many people carrying a damaged gene would be detected that it is hard to know what to do with the information—or whether it was worth gathering in the first place.

There is controversy about the whole idea of mass screening. One problem is that many tests are not absolute. Because the same disease may arise as a result of mutations in different parts of the gene (or because the damaged gene is associated with different segments of DNA in different families) tests often miss a proportion of carriers. For cystic fibrosis, most miss about one British carrier in ten—and do even worse in other parts of Europe. In Turkey or Israel the standard British test would identify only a third of those carrying the gene. Screening can never provide reassurance that someone is not carrying the gene: the best it can do is tell them that they are a carrier, or that the test is inconclusive. Uncertainty of this kind can cause anxiety and distress.

Some couples believe, wrongly, that marriages in which only one partner has the mutation are in danger of producing an affected child. As a result, many centers only describe the result as positive if both partners are carriers. There are also unpleasant reminders of the past. In a recent survey, not one German doctor was willing to accept such a screening program, even if it was absolutely reliable and the results were kept private. Because of the way in which the issue is entangled with the abortion debate the American Cystic Fibrosis Foundation concentrates on appealing for funds for a cure rather than a test. No population screen is planned for the U.S.A. Mass testing would be expensive, too. It might cost as much as fifty thousand pounds—seventy-five thousand dollars—for every case of cystic fibrosis prevented in Britain (although this is much less than the cost of lifelong treatment).

The new biology has the potential to be a more positive force than it is at present. Understanding a disease may be the first step toward treating it. In principle, genetics can already do a lot more than just test for abnormalities. It can already treat some inborn diseases. Sometimes, the treatment is simple. Children born with phenylketonuria, an inability to cope with a certain amino acid, used to die young. However, all that is needed to alleviate its symptoms is a diet which lacks the amino acid. Other therapies for inherited illness are more complicated, but work just as well—for example, injection of a blood-clotting factor cures the symptoms of hemophilia. Sometimes, more drastic action is needed. Cystic fibrosis is lethal because the lungs fill with mucus. A heart-lung transplant can help. This involves a risk, but offers the chance of a reasonably normal life. The discovery of the cystic fibrosis gene is already beginning to help in treating the disease. The virus of the common cold can be persuaded to incorporate a working version of the gene which has gone wrong in cystic fibrosis. Spraying this into the nose of patients reduces the severity of the affliction—so far only slightly; but there is hope of rapid progress.

These treatments deal only with the symptoms of genetic damage—which is, of course, exactly what medicine does for most diseases. Gene therapy gives hope of a cure. It involves replacing a faulty section of

DNA. Copies made in the laboratory are inserted into a living cell with the help of a virus. Working genes can even be shot into cultured cells by firing DNA from a tiny gun. In principle, this approach is a powerful one. In the early days—a decade ago—there were great hopes that it would revolutionize the treatment of inborn disease. Although the triumph of gene therapy has been just around the corner since it began it has not yet fulfilled its early promise. In 1994, more than fifty trials are planned or under way and gene therapy may soon begin to achieve its goals.

There have been a few successes already. Severe combined immunodeficiency (or SCID) is an inherited failure of the immune system. Children with the condition are sometimes kept in a plastic bubble to reduce the chances of infection and are given bone marrow transplants to boost their defenses. A specific enzyme is missing. Cells in culture which lack the enzyme have been "cured" by inserting the appropriate DNA. In 1990 two children were treated with such engineered cells. They are still alive and one is even going to school. As they were simultaneously treated with extracts of the enzyme taken from animals it is not yet certain that their improved health is due directly to gene therapy.

Work is well advanced on therapy for more common diseases, too. There is the prospect of introducing into living cells the gene for one of the clotting factors which fail in hemophilia. If the scheme succeeds, it may be the basis of a treatment. Some families have a gene which interferes with the removal of fat from the blood. Those affected are at risk of coronary heart disease. Children with two copies of the damaged gene have little chance of surviving to middle age. One child has been treated with cells into which a working copy has been inserted. There are encouraging signs that the treatment is helping to clear fat from his blood.

There are great hopes for gene therapy of this kind. However, some of the most frequent inherited diseases are going to be difficult to treat. To cure sickle-cell anemia would involve targeting tiny numbers of cells deep in the bone marrow, as it is these—and not the red blood cells themselves—which produce the faulty hemoglobin.

There are other ways in which molecular biology might be used to treat disease. Cells engineered to carry genes which destroy cancer cells or cause them to stop dividing could be used in therapy. It might be possible to introduce genes which stimulate the immune system's defenses against cancer into cancer cells themselves, providing them with the seeds of their own destruction. Another trick, which has worked on animals, is to insert genes into cancer cells which make them susceptible to particular drugs. This works best against brain tumors; in the brain, almost no normal cells divide, so that only the cancer cells take up the foreign gene. Once the DNA sequence of a faulty gene has been established, there is the prospect of making an "anti-sense" nucleic acid which binds to the genetic machinery and blocks it. Perhaps it will be possible to switch off genes (such as cancer genes) which have gone wrong. Anti-sense therapy may prove to be as important a medical breakthrough as was penicillin.

Even if these hopes for therapy fail, there is hope for great improvements in diagnosis. New probes will detect mutations (such as those which lead to cancer) long before the symptoms of disease first appear. Cancer cells often develop unusual antigens on their surface as new genes are switched on. Once the DNA involved has been identified it should be possible to work out the shape of the protein which produces the antigen. A matching protein could then be made which sticks only to the cancer cell. If a drug is attached, it will bind exclusively to such cells. Much more powerful chemotherapy will then be possible, as there will no longer be the risk of poisoning normal cells as well.

Biology might do even more: in theory it could be used to treat generations yet unborn. In mice genes can be inserted into egg cells so that they are passed on to future generations. The germ line, as it is known, has been changed. Such "transgenic mice" are valuable research tools. If genes for a human disease are introduced they can be used to study its symptoms (although these may differ from those found in humans themselves) and the mice may even be used to test drugs which could later be used in treatment. There are now transgenic mice for sickle-cell anemia and other inherited illnesses. There is a transgenic pig, perhaps the first of many, which contains some of the genes for human cell surface variation. This means that its tissues—heart or kidney, say—which are about the right size for transplantation, are more acceptable to a human patient than they otherwise would be. The pig looks, of course, just like a pig, but, to our immune system, its organs resemble those of a human being. A counterfeit human heart has not yet been used for transplantation, but it may soon be. Human germ line manipulation has never been tried, but there is no great practical reason why it should not. If it succeeded, those with a disease and their descendants would both be helped.

To some, all this is the first step toward Frankenstein. Many of the fears are exaggerated. Most treatments of inherited disease are not very different from those in the rest of medicine. Surely no one worries about treating phenylketonuria with a special diet or hemophilia with factor VIII. And a society which accepts a heart-lung transplant for a child with cystic fibrosis cannot deny it the right to have the symptoms treated at their source with a working gene. All that is different is the level of intervention—the DNA itself rather than what it produces.

There are clear rules which apply to every medical therapy. Everyone has rights to their own body and can decide whether or not to accept treatment. Exactly the same kind of reasoning can be applied to genes. Replacing damaged DNA is not much different from replacing a damaged kidney and the same choices must be made by the person who receives it (or by their parents). Changing genes in sperm or egg is different. It alters the inheritance of someone who has no choice as, in a few years, the manipulated gene will be carried not just by those who agreed to it but by their descendants. On this and other grounds germ line therapy is generally thought to be unacceptable. There is even a move afoot to add to the Universal Declaration of Human Rights a statement that everyone has the right to a genetic constitution which has not been changed without their consent.

The moral problems which arise from interfering with DNA are less immediate than another and more subtle ethical issue. This is the dilemma of knowledge: the fact that genetics can tell us things we may not wish to know. Huntington's Disease (see p. 69) has an unusual pattern of inheritance. Those with a single copy show its effects, but often not until middle age. Anyone with a parent who has the disease has a one in two risk of having inherited the gene. Because of the delay in signs appearing such people are left in uncertainty as to their fate. Symptoms usually display themselves in a patient's thirties or forties as a general restlessness followed by involuntary movements and ending in paralysis and death, usually within twenty years of diagnosis.

Although the gene itself has only just been tracked down, there are changes in the DNA close to it which have already been used to test whether someone has inherited the disease long before the first signs appear. Because the changes are also found in some normal people, relatives—grandparents, parents and even cousins—of those at risk must be tested to establish what variants are associated with the Huntington's gene within a particular family. This means that some people who did not previously appreciate their risk of carrying the gene may learn of their fate.

Those with an affected parent are at risk of passing the affliction to their own offspring. Sometimes they choose to have their pregnancies tested (and to end them if the fetus is affected) even if they themselves have no symptoms. Should the fetus be carrying the gene, then the parent at risk will know that he or she carries it and may soon become ill.

In Britain, only a few hundred of the thousands of people who know themselves to be at risk of carrying the Huntington's gene have come forward since the test was first offered in 1987. Perhaps the knowledge about one's fate is too hard to bear, so that people choose to live with risk rather than with certainty. The new genetics is making certainty, unattractive though it may turn out to be, an option for more and more of us.

Huntington's Disease illustrates the ethical difficulties of knowing about our genes. Soon this knowledge will be available to many more people, should they choose to have it. The great killers of the modern world, heart disease and cancer, are strongly influenced by genes. Before long, many could learn the probable date of their death. What this may do to society has scarcely been considered (although it might make an interesting plot for a novel set in the future).

Knowledge brings more mundane problems, too. Insurance—of any kind—is a mechanism for diffusing risk. Buying a policy means that the cost of an accident is diluted by sharing it with those who have done the same thing but never make a claim. House or car coverage is based on a knowledge of risk. Those who enjoy driving drunk or keeping gold bars under the bed pay more and do not complain (much) that their life-style forces them to do so. But what about health insurance? In the United States (and increasingly in Britain) access to medical care is limited by the ability to pay. Fifteen million Americans pay for their own health coverage. For most of the others the insurance premiums are paid by their employer (although thirty-five million have no health insurance at all). Anyone buying a policy has to disclose any medical problems of which they are aware. Already about a third of applications are denied and for the remainder all "pre-existing conditions"—reported or not—are excluded.

Genetic testing raises some huge questions. Should the insurance company have the right to demand the results of a test to help them decide how much to charge or even whether to insure at all? Is a damaged gene a pre-existing condition? After all, everyone must die: and genetics can do no more than tell some people when this might happen. But health insurance depends on spreading risk. Genetics may be a terminal blow. It erodes ignorance of future disease. No one will play with a gambler who knows all his opponent's cards and no one will pay for health coverage when they are pretty certain that they will live to a ripe old age (and will not need it) or when the company knows that an expensive illness is programmed into the genes and charges accordingly. Insurance already suffers from the fact that people at high risk are more likely to buy a policy. There may be a war of cost escalation which ends with only risk-prone people paying for health insurance. The Director of Communication for the American Council for Life Insurance has said that he "wishes that genetic technology had never been developed."

Already, the companies are refusing to insure those doomed to Huntington's Disease—and there are many other ailments of the same kind. Denying coverage is no empty threat. Recently, a woman in charge of a fragile X screening program (see p. 77) in the United States was denied insurance because her children had symptoms of the disease, although she had none. In another case, the insurer agreed to pay for a fetus to be tested to see whether it had cystic fibrosis, but only if the parents agreed to have an abortion if the test was positive.

Genetic information means that insurance will no longer be blind. In a commercial health market there will be good and bad buys. If the employer pays the bill there will be pressure not to hire someone whom the screening program shows to be in jeopardy. All this is an argument for a national health service, which diffuses individual risk among the whole population. Perhaps health care will revert to a police rather than security-guard role, with the state accepting that all must pay equally, although some are in more danger than others.

Many people are concerned about what genetics is doing to the future. Are we in danger of producing a race of Frankensteins? Victor Frankenstein himself, who created the nameless monster in Mary Shelley's Gothic tale, considers at the end of the novel whether to produce a mate for his chimera but dismisses the idea because ". . . one of the first results of these sympathies for which the daemon thirsted would be children, and a race of devils would be propagated upon the earth who might make the very existence of the species of man precarious and full of terror. Had I the right, for my own benefit, to inflict the curse upon everlasting generations?"

Genetics is often seen as a threat and to interfere with our inheritance as a curse on the future. In fact, the new biology has brought little but benefit. Ironically enough, Mary Shelley was pregnant when writing *Frankenstein* and produced a child which died when a few months old. She herself suffered from a severe depression, as did many of her relatives. Today's genetics might have helped to understand her child's illness and her own mental state. Deliberately to interfere with our genes will have some effects on the "everlasting generations," but these are likely to be much less important than are some of the evolutionary changes which are happening without our realizing. What these are I will consider in the last chapter.

16 | The Evolution of Utopia

ONE REASON science fiction is so boring is that it is nearly all the same. Although the monsters may differ, the plots do not. The same is true for most imaginary Utopias. From *The War of the Worlds* to *Planet of the Apes* an alien life form appears, masters the human race, and meets its doom because of its own biological failings. Most novels of the future ignore one of the few predictable things about evolution, which is its unpredictability. No dinosaur could have guessed that descendants of the shrew-like creatures playing at its feet would soon replace it: and the chimpanzees who outnumbered humans a hundred thousand years ago would be depressed to see that their relatives are now so abundant while their descendants are an endangered species.

Evolution always builds on its weaknesses, rather than making a fresh start. The lack of a grand plan is what makes life so adaptable and humans—the greatest opportunists of all—so successful. Life's utilitarian approach also means that speculating about the future of evolution is a risky thing to do as it is difficult to guess what a pragmatist will do next. As Hegel put it, the greatest lesson of history is that no one ever learns the lesson of history.

In this, the last chapter, I will take the risk of making a genetical forecast. I am certainly not the first to do so. Many of the best-known Utopian novels trace their visions of the future directly to Darwin. Samuel Butler, author of *Erewhon*, shared an education—Shrewsbury School and Cambridge—with Darwin, and was himself a keen evolutionist (albeit an anti-Darwinian). Aldous Huxley's *Brave New World* owes much of its plot to his biological brother Julian and to their grandfather Thomas Henry Huxley, who was known as "Darwin's Bulldog" because of his fierce defense of his hero. H. G. Wells—whose Utopia appeared in *The Shape of Things to Come*—himself wrote a biological textbook with Julian Huxley; and, as we have seen, George Bernard Shaw, author of *Back to Methuselah*, was a follower of Galton and appeared on public platforms with him.

Sometimes the link between the utopian novelists and eugenics is embarrassingly clear. Shaw felt that "if we desire a certain type of civilization we must exterminate the sort of people who do not fit into it." H. G. Wells shared his views. In his scientific vision of the world to come, the (now obscure) *Anticipations of the Reaction of Scientific Progress upon Human Life and Thought*, published in 1901, he wrote in favor of euthanasia for "the weak and sensual" and genocide for "the dingy white and yellow people who do not come into the needs of efficiency." Many of the most famous Utopias would not have been very comfortable places for the people forced to live in them.

The whole of this book has been a tale of how humanity evolved by the same rules as those which propel less pretentious creatures. Humans are, of course, more than just apes writ large. We have two unique attributes: knowing the past and planning the future. Both talents guarantee that the outlook for humankind will depend on a lot more than genes. Nevertheless, it should be possible to make some guesses from the biological past as to what the evolutionary forecast might be. One pessimistic but probably accurate prediction is that it means extinction. Although about one person in twenty who has ever lived is alive today, only about one in a thousand of the different kinds of animal and plant has survived. Our species is in its adolescence, at about a hundred and fifty thousand years old (compared to several times this for those of our relatives whose fossil record is good enough to guess their age). Its demise is, one hopes, a long way away; and we can at least reflect about what might happen before then.

The rules which drive evolution are simple and are unlikely to change. They involve the appearance of new genes by mutation, natural selection, and random changes as some genes, by chance, fail to be passed on. To speculate about what will happen to each of these processes is to make predictions about human evolution. Will this biological Utopia be anything like its fictional equivalents (as I hope it will not); will we continue to evolve as rapidly as we have since our beginnings, or is human evolution at an end?

Human beings have interfered quite unknowingly with their biological heritage since they first appeared on earth. As we saw in earlier chapters, stone tools, agriculture and private property all had an effect on society and in turn on evolution. Many people are concerned that the next phase of human history will be one in which genetics makes deliberate plans for the biological future. This is expecting too much of science. Inadvertent change—evolution by mistake—is much more likely to be important than is any conscious attempt to modify biology.

Even the most determined efforts of doctors, genetic counselors or gene therapists will have only a small effect on future generations. About one American child in two thousand five hundred is born with cystic fibrosis—but a hundred times as many carry the gene without knowing it. Molecular biology makes it possible to advise people of their condition and perhaps, one day, will provide a cure. Even today's imperfect treatment means that the number of affected children surviving to reproductive age will double in the next thirty years. Nobody knows what the balance will be; whether the fact that more of those with cystic fibrosis pass on the gene will be outweighed by a decrease in the number of sufferers as genetical advice allows parents to plan their reproduction. Many people with phenylketonuria have had children. There was once strong social pressure against those with inborn disease marrying. In the 1950s only a minority of achondroplastic dwarfs found a spouse, but in the United States more than eighty percent of them are now married, often to another dwarf. No doubt, many genes which once disappeared quickly as their carriers died or remained single will now persist.

This is unlikely to have much influence on the biological future. Most inborn diseases which are susceptible to treatment or to prenatal diagnosis are recessive, so that there are hundreds of times as many copies of the gene in healthy people as in sufferers. As everyone carries several hidden recessive mutations there is little prospect that medicine will pollute a once pure human gene pool by allowing a few more copies to survive.

Many inherited diseases appear anew each generation by mutation. Is the evolutionary future in danger because of an increase in the mutation rate?

There are real concerns that modern civilization—with its dubious benefits of nuclear radiation and poisonous chemicals—will lead to a dramatic increase in the number of mutations. In many science fiction worlds this is, in a few short generations, enough to degrade the human race. The obvious threats, including manmade radiation and chemicals, have a smaller effect than do natural mutagens such as radon gas leaking from granite (see p. 79). The Sellafield nuclear power station in the north of England is the most polluting in the western world and the North Sea the most radioactive body of water. The name of the station has itself mutated from Calder Hall to Windscale to Sellafield in an unavailing attempt to calm public suspicion. Yet, compared to other sources of radiation, its effects are minor. Avid consumers of shellfish collected near the discharge pipe receive about as much excess radiation as do those who fly from London to Los Angeles and back four times a year and are exposed to cosmic radiation as a result.

A more subtle transformation is having a dramatic effect on the mutation rate. In the western world at least a change in the age at which people have children means that the number of new mutations will probably drop.

The rate of mutation goes up greatly with age. The control of infection means that most people now live for far longer than in earlier times. Mutation can hence take its toll on a much higher proportion of the population. This is very obvious when looking at such changes in body cells, including those which give rise to cancer. The cancer epidemic in the modern world is largely confined to older people. A shift in the pattern of survival has had effects on genes as they reside in body cells.

Cells which give rise to sperm or egg are also exposed to the destructive effects of old age. Older parents are more likely to have genetically damaged children than are those who reproduce early. Any change in the age of reproduction will hence have an effect on the mutation rate. If the number of elderly parents goes up, there will be more mutations; if it decreases, there will be fewer. Social progress has led to just such a shift. The general picture, worldwide—a picture which applies to most of the third world as much as to Britain and the United States—is simple and a little surprising.

Before the improvements in public health over the past few centuries most children died young. Mothers started having babies when they were themselves youthful and continued to have them until they were biologically unable to do so, perhaps twenty-five years later. As infant mortality drops there is less pressure to have children as an insurance against one's own old age. People prefer to have smaller families. The availability of contraception means that parents can choose to delay their first child—sometimes until their mid-twenties—

but then complete their families quickly. This means that most people stop having children soon after they have started. As a result, the number of older mothers and fathers goes down as social conditions improve.

The effect is obvious in postwar Europe. In countries such as Poland and Switzerland the proportion of mothers aged more than thirty-five —the group most at risk of mutation—dropped from around twenty percent in 1950 to less than five percent in 1985 and is still falling. Less than one mother in fifty in what was East Germany is more than thirty-five years old. The effect is particularly striking in Ireland. The influence of the Church—and the fact that many young men spent a period working overseas—meant that until a few years ago the main means of birth control was self-denial. Most Irish people did not marry until their late twenties, or even later, and until recently there were more than twice as many mothers over thirty-five in Ireland than anywhere else in Europe. The number is now dropping rapidly (although it is still well above the European average). In Britain and the Scandinavian countries there was a slight reversal of the trend toward earlier reproduction in the mid-1970s, with the number of mothers over thirty-five increasing slightly today from its low point of around one in twenty.

All this means that there are fewer old mothers than in much of the evolutionary past. Fathers, too, are getting more youthful. This is bound to have an effect on the mutation rate. Down's Syndrome is ten times more common in mothers over forty-five than those under twenty-one. It is three times more common in Pakistan (which has almost no family planning) than in Britain, largely because Pakistani mothers are older than their British equivalents. Looking at things from the male point of view, in Britain the mutation rate in men is about one and a half times that expected if all fathers were less than thirty, but in Pakistan it is still three times this low figure. At the moment, at least, it looks as if the human mutation rate is on the way down. Whether this trend will continue is not known, but it does put fears about a new race of mutated monsters into context.

If mutation is the fuel of evolution, natural selection is its engine. As selection is a more elusive process than mutation it is more difficult to forecast what its future might be. Nature is always liable to come up— as it has so often before—with a nasty surprise with which natural selection must cope. The emergence of the AIDS virus shows that there is an eternal risk of this happening again. However, in the western world at least some of the greatest selective challenges have gone, because of the control of infectious disease.

Once a disease has disappeared (as so many have) the fate of the genes involved in combating it will change. Cypriots carry the inherited anemia beta-thalassemia because the gene once defended their ancestors against malaria. Malaria has now disappeared from Cyprus—as, in time, will thalassemia, with the incidence of carriers likely to drop by as much as one percent per generation from its present level of seventeen percent. In time, and given success—still far distant—in the fight against malaria, the same will happen to the dozens of other genes involved in resisting it elsewhere in the world. Perhaps such genes will, in time, remain only as mute witnesses to their evolutionary past.

Civilization brings its own afflictions. Coronary heart disease and diabetes are illnesses of diet, of fat and sugar. Those with certain genes are more likely to be affected than are others. Perhaps their inheritance was advantageous when the food supply was unpredictable. Only with a constant rich diet did it become dangerous. The change in diet has already altered the pattern of natural selection on the Pacific island of Nauru (see p. 188). Now that people are modifying what they eat the risks may go down again, altering selection once more. The evolutionary future depends on an environmental shift. As the action of so many genes depends on the environment in which they find themselves changes in life-style may have as much of an effect on evolution as do changes in the DNA itself.

The history of one inherited character, the weight of babies, shows just how effective improved conditions can be in reducing the action of natural selection. Birth weight shows the advantage of being average. Not surprisingly, underweight babies survive less well than do others. What is more remarkable is that babies heavier than average are also more likely to die in the first few weeks. In the 1930s, about half the babies who died in their first year did so because they were above or below the ideal weight. A difference of just one pound had a major effect on survival. Since some of the variation in this character is genetic, natural selection was at work against genes for high or low birth weight as it had been, no doubt, since our species began.

Now, such selection is disappearing. Improved health care means that only very underweight babies, or those much larger than average, are at risk. The intensity of natural selection went down by two thirds between 1954 and 1985. Nowadays there is very little risk in being a baby of even two pounds above or below the mean birth weight. What must once have been one of the strongest agents of selection (acting as it does before the age of reproduction) seems to be on the way out. Another subtle effect of improved child care will be to change the ratio of the sexes at the age when most people choose a mate. There are slightly more boys than girls at birth. Boys were once less likely to survive the hazards of childhood, which meant that there was almost exactly the same number of each sex in those significant years, the late teens. Now, boys survive almost as well as girls do, so that in the future there will be a slight but noticeable excess of young men looking for a partner.

There are more subtle ways of looking at the future of selection than just multiplying examples of how it works. Natural selection can only act on differences. If everyone survived to adulthood, found a partner and had the same number of children there would be almost no chance for it to operate. We do not need to know what genes selection is working on to see how important it might be. Looking at changes in the patterns of birth and death reveals a lot about its actions in the past and in the future.

In affluent countries, the differences between families in how many people survive have decreased. This means that there is less opportunity for natural selection. Ten thousand years ago, the struggle for existence really meant something. Skeletons from cave cemeteries (such as that of the Indian farmers preserved in the excavated cemetery at Libben, Ohio) show that few people lived to be more than twenty. If ancient fertility was anything like that found in modern tribal groups each female had about eight children, most of whom died young. For nine tenths of human evolution, society was like a village school, with lots of infants, plenty of teenagers and a few—probably harassed—adult survivors. Almost every death was potential raw material for selection as it involved someone young enough to have had a hope of passing on their genes. Nowadays, things have changed. Ninety-eight out of every hundred newborn British babies live to the age of fifteen, so that selection acting through childhood deaths (once its main mode of operation) has almost disappeared.

Modern India is a microcosm of how selection is losing its chance to mold the human condition. The country contains a wide range of lifestyles, from tribal hill-peoples to affluent urbanites. As a result it contains within itself a history of social change over the past several thousand years. Combining information from the various groups on differences between individuals in the chance of survival and the number of children shows that natural selection has lost eighty percent of its potential in middle-class town-dwellers compared to their fellow citizens who follow a tribal way of life.

There have been changes in the balance of birth and death which have other effects on the opportunity for natural selection. Few modern peoples are as fertile as they once were. The Hutterites in North America wish for the largest possible family for religious reasons but even they, living in a healthy society, rarely have more than ten children. For most of human history it seems that people had as many children as is biologically possible. Only recently has that number begun to decrease.

Only in the past few years, though, have humans lived as long as they are able. In the West, average life expectancy has nearly doubled over the past century. For the first time in history, most people die old; perhaps as old as biology allows. Life expectancy has risen from forty-seven to seventy-five years since 1900. Progress has now stopped, for some social classes at least. In the U.S.A. in 1979, a white woman of sixty-five could expect to live for another eighteen and a half years. In 1991 the figure was exactly the same. Even if all infectious diseases and all accidental deaths could be eliminated, average life expectancy in the western world would now go up by only a couple of years. There is still room for progress in the average length of life because of class differences in health. A baby born to an unskilled worker in Britain can expect to live for eight years less than one born to a professional person, a difference which, to our national shame, is actually increasing. However, the prospects for any dramatic improvement in longevity are dim. George Bernard Shaw was wrong. There is not much chance that we will go back to Methuselah.

This is important for the evolutionary future. The increase in the number of old people means that more people die for genetical reasons than in earlier times (largely because fewer are killed violently or by infection). Paradoxically, it also means that selection is weaker. The genes that kill are now those for cancer or heart disease, which act late in life. Those who die have already reproduced, passing on their fatal inheritance. Natural selection is much less powerful when acting on genes such as these than it is on those which alter the chances of survival before their carriers have children.

The new pattern of human existence (with fewer children than ever before but most people lasting until the biological clock runs down) emerged only about twenty human generations ago compared to the six thousand or so since we first appeared on earth. It means that natural selection has changed the way it works. What there is nowadays acts more on fertility than on survival.

Differences in fertility among families, and the opportunity which they give for selection, shot up as birth control became popular. The upper classes adopted the idea well before the lower orders. The French aristocracy caught on first and reduced the number of children per marriage from six to two in only a hundred years. In Victorian times, variation in fertility was striking. Mr. Quiverful, in Trollope's Barchester novels, had a dozen children at a time when the other clergymen were discreetly limiting their families to two or three. Now that birth control is widespread the differences between families have dropped again, but selection working through variation in the number of children born is still, for the first time in history, greater than that working on the number that survive. The evolutionary fate of our genes depends more on the number of children we choose to have than on their chances of staying alive.

Nearly all the best understood forces of selection—disease, climate or starvation—act on survival rather than on fertility. The shift in the balance of the two may bring in new and unpredictable evolutionary forces. Perhaps the age of reproduction will become important, as those who mature young can have more children. There has been a drop in the age at which girls become sexually mature. In opposition to this trend, western women now marry five years later than they did half a century ago. Any inherited tendency to marry earlier or later (or to limit family size) could become a potent agent of evolution.

What this will do to the biological future is hard to say. A good general rule in evolution is that nobody gets a free lunch: success in one walk of life must be paid for by failure in another. Experiments on fruit flies suggest that a shift away from the ability to survive toward selection on fertility involves a trade-off. Flies which produce lots of eggs die young. Perhaps, in time, the same may happen to humans.

Whatever the long-term outcome of natural selection, there is no reason to think that it will change its tactics. Rather than making a dramatic new start by designing the ideal solution for a particular problem—a design which may not be ideal for very long—it will build on existing imperfections. History gives little reason to hope that selection will act as the agent of human perfectibility. It may direct the future, but will never make humanity superhuman.

The number of new mutations and the intensity of natural selection are both declining. This certainly does not mean that evolution is over. There is another change in modern society which is bound to influence our biological prospects. It is one which most people scarcely consider. It has to do with the geography of mating.

For most of history, everyone more or less had to marry the girl (or the boy) next door, because they had no choice. Society was based on small bands or isolated villages and marriages were within the group. In many places, populations were stable and quite inbred. Almost nobody moved. The genes of American Indians, drowned in a peat bog in Florida, display the effect clearly. The DNA in the preserved brains of people who died a thousand years apart shows that their genes are almost the same. There was little migration and the Indians had no option but to marry their relatives.

This pattern persisted in the West until recently and still exists in many parts of the world. In some places it is changing quickly. An increase in mating outside the local group is the most dramatic change in the developed world's evolutionary history. The effect is getting stronger and stronger. The influence of outbreeding on genetical health will outweigh anything that medicine is able to do.

Some societies once encouraged mating with outsiders. In the Ottoman Empire, talented people were produced by promoting marriages between people from different nations. Their children were seen as "like the fruit of a union of two different species of tree; large and filled with liquid, like a princely pearl." In South America after the arrival of the Spaniards there was what was described as "a conquest of the women" by the invaders. Paraguay—the site of Elisabeth Nietzsche's failed genetical experiment—became known as the Paradise of Mohammed, and every Spaniard had an average of twenty or thirty Indian women. The Governor excused this, saying that "the service rendered to God in producing mestizos [children of mixed race who were raised as Christians] is greater than the sin committed by the same act." It is pleasant to believe that increased mating with outsiders had something to do with concern for genetical health, but lust is a more likely explanation.

Outbreeding is not usually due to deliberate policy. Much of it arises, like so many of the biological events which shaped the human condition, as a by-product of social change. Cities and transport both played a part as both provided a larger pool of potential mates than was available in the days of rural solitude.

In parts of Europe in the early part of this century marriages between relatives were still common. In the Aeolian Islands off the coast of Italy in the 1920s a quarter of marriages were between first or second cousins. This figure has dropped to about one in fifty (and in Italy as a whole it is now less than one percent). Britain has always been more outbred than most of Europe, but the effects of increased outbreeding can be seen here, too, with a striking drop in cousin marriages since Victorian times.

Elsewhere, the picture is not so simple. Some non-European societies promote marriages between relatives for economic reasons. They are still frequent in Indian villages, where up to half of all unions may be between cousins, or between uncle and niece. Indeed, among Pakistani immigrants to Britain the incidence of first cousin marriages is greater than in their native land, perhaps because of social isolation. Nearly half of British Pakistanis of reproductive age are married to a cousin, a rate of cousin marriage—and hence of inborn disease—higher than that among their own parents. Any move toward more integration and less inbreeding in the next generation will have a marked effect on genetic health.

One way of illustrating changes in breeding pattern is to use a crude but effective measure of how closely related our ancestors may have been. All that one needs to know is how far apart they were born. If they come from the same village they may well be relatives, but if they were born hundreds of miles apart this is much less likely. For nearly everyone reading this book the distance between the places where they and their own partner were born is greater than that separating their parents' birthplaces. In turn, today's fathers and mothers were almost certainly born farther apart than were their own parents. In nineteenth-century Oxfordshire the distance between the birthplaces of marriage partners was less than ten miles. Now it is more than fifty. In the United States it is several hundred, so that most American couples are completely unrelated. All this shows how quickly the world's populations are beginning to mix together.

It will take a long time before the mixing is complete. One estimate is that it will take as much as five hundred years to even out the genetic differences between England and Scotland—and perhaps even longer to get rid of their cultural contrasts. Even if global homogeneity is a long way away, increased movement will certainly have a biological effect. No longer will large numbers of children be born who have two copies of defective genes because their parents are related. Think of a mating between an African slave and a white slaveowner in nineteenth-century America. Each had a chance of carrying one copy of certain defective genes. The most common damaged gene in whites is the one for cystic fibrosis, in blacks that for sickle-cell anemia. Only if a child inherits two copies of either of these will it suffer from an inborn disease. Because cystic fibrosis is unknown in Africans and sickle cell in whites the child of a black-white mating is safe from both illnesses.

This effect can be a strong one. In many societies in the modern world there are immigrant communities which are beginning to merge with the people already there. Imagine that ten percent of the population of Britain were to immigrate from West Africa (where about one person in fifteen is a carrier of the gene for sickle-cell anemia) and to mate freely with the locals. The number of sickle-cell carriers in the new mixed British population would go up by seven times. However, the incidence of sickle-cell *disease*—which demands two copies of the damaged gene, one from each parent—would drop by ninety percent compared to the previous situation in the two groups considered together. This is because many children would be born to parents from the two different peoples, one of whom—the native British partner—does not carry the sickle-cell gene. There would also be an effect on the indigenous British disease, cystic fibrosis, whose incidence would drop by about a sixth. Although this model of race mixing is simplistic it is not completely unreasonable. In Britain now, about one marriage in thirty is between two people of non-European origin; but a third as many is between a non-European and someone whose ancestors were born in the British Isles. In the United States, the process of blending is well advanced. The genes of black Americans show that there has been mating between Americans of African and of European origin (p. 38) for several hundred years. This has certainly improved the genetic health of their offspring.

This change in mating patterns may mark the beginning of a new age of genetic well-being. Increased outbreeding inevitably means that recessive genes will be partnered by a normal copy which masks their effects. This is enough to dwarf the efforts of scientists to improve genetic health.

In time the newly mixed populations of the world will reach a new equilibrium. Many of the recessive genes hidden in the descendants of mixed marriages will reappear. This will take thousands of years. There is little doubt that the most important event in recent human evolution was the invention of the bicycle.

Patterns of mating and the genetic future will also be influenced by the dramatically different rates of population growth in different parts of the world. The overwhelming evolutionary fact of the past three centuries has been the population explosion. By the time the Pilgrim Fathers arrived in America, the population of the world was only about twice that on the first Christmas Day. Since then, human numbers have increased at an astonishing rate. In the living world as a whole, evolutionary change is a much slower process than are shifts in geographical distribution and in abundance. Many populations—and most species—go extinct before they have a chance to react to an environmental challenge. The threat of disaster as a result of overpopulation may mean that all speculation about the genetical future is irrelevant.

There is always a delay between an improvement in health care—and the subsequent increase in population size—and a decrease in the number of children which parents choose to have. This delay explains the recent explosive growth of the world's population, which has doubled since 1950 to its present level of five thousand million. It is projected by the United Nations to double again by the year 2050 and to rise to thirteen thousand million before the end of the next century. This rate of growth is equivalent to adding a country the size of Mexico each year.

There have been great contrasts in the way in which different societies have changed the pattern of reproduction. The West moved to the new way of life earlier than did the third world. This means that population growth is now far more rapid in certain parts of the world, such as Africa, than elsewhere. The United Nations estimates that more than ninety percent of the global population rise will be in these regions. Africa will grow particularly rapidly as it shows little sign of a drop in birth rate. The number of children per woman in East Asia decreased from 6.1 to 2.7 between 1960 and 1990; but in Africa the figures for those years were 6.6 and 6.2. Some estimate that a third of the globe's population will be of African origin by the year 2050. Recent fears about the effects of epidemic AIDS cast some doubt on this figure.

It is certain that some populations will continue to grow more rapidly than others. As the groups involved are genetically distinct—Africans, for example, contain unique genes for disease resistance and are generally more diverse—this in itself represents an evolutionary change. In the past (as after the agricultural revolution) population growth which followed an economic change led to migration. Although there are political barriers to mass movement in the modern world these are unlikely to work for long. It may be, then, that the future Utopians will mainly be black.

Nevertheless, most social changes seem to be conspiring to slow down human evolution. Mutation, selection, and random change have all lost some of their effectiveness in the past few centuries. All this means that the biology of the future will not be very different from that of the past. It may even be that economic advance and medical progress mean that humans are almost at the end of their evolutionary road, that we are as near to our biological Utopia as we are ever likely to get. Fortunately, no one reading this book will be around to see if I am right.

Selected
Bibliography

TRYING TO KEEP UP with the scientific literature is like running up a down escalator. However much one reads, more and more is published. Finally one is, inevitably, forced to give up from mere exhaustion. As genetics is now moving so quickly it is necessary to sprint upward to stay in the same place. Although the fundamentals of the subject have not changed much in the past few years, many of the discoveries described here—the fact that Huntington's Disease is due to multiplication of short sequences of DNA, the so-called "gay gene," the transgenic pig and the maritally choosy Hutterites—were made in the 1990s, some in the few months before this edition appeared.

Papers at the cutting edge of any field tend to go out of date quite quickly. With the exception of a few key research papers, I have therefore not tried to refer to all the sources used in writing *The Language of Genes*. For those with access to a good library, the British journal *Nature* and its American equivalent *Science* publish almost every week new discoveries in human genetics and evolution, accompanied by review articles which put the findings into a wider context. Scientists used to read them from the back (where the job advertisements are). It is a sign of the excitement of modern genetics—and, perhaps, of the decline of the academic job market—that more and more of my colleagues now study *Science* and *Nature* as they are written, from the first to the last page.

There are many specialist journals in human genetics. *The American Journal of Human Genetics* is the pacesetter and has reviews and editorials as well as its scientific papers. *The Annals of Human Genetics* and *Nature Genetics* are also important sources. *The American Journal of Physical Anthropology* approaches human evolution from a less genetical angle.

Scientific American provides up-to-date and well-presented information on genetics and evolution, published almost as it happens. Sometimes, even the newspapers get it right.

There are a number of outstanding books about genetics and evolution. One of the best is *The Rise and Fall of the Third Chimpanzee* by Jared Diamond (1991, HarperCollins). His title comes from the discovery that chimps and humans share most of their DNA sequences; and Diamond uses this to build a fascinating and engagingly written history of what we can learn about ourselves from our living relatives. A more sedate account of human genetics is in *The Code of Codes: Scientific and Social Issues in the Human Genome Project*, edited by D. J. Kevles and L. Hood (Harvard University Press, 1992). Several distinguished authors give a full account of the new technology and what it means—not forgetting its implications for law and for ethics.

For a more pessimistic view of its possible dangers—which may perhaps have been downplayed in my own pages—there is *Perilous Knowledge: The Human Genome Project and Its Implications* by Tom Wilkie (Faber and Faber, 1993). Richard Lewontin's book lives up to its title *Biology as Ideology* (Penguin Books, 1993) with a lively and provocative onslaught directed at the self-righteousness which has pervaded much of human genetics since its beginning. The confidence of geneticists then was quite misplaced and Lewontin argues that the same is often true today.

A more technical (but still eminently readable) account of human genetics is in David Weatherall's *The New Genetics and Clinical Practice* (3d edition, 1991, Oxford University Press). Those not put off by scientific terminology—and genetics is plagued with jargon—should try *Recombinant DNA* by J. D. Watson, M. Gilman, J. Witkowksi and M. Zoller (2d edition, 1992, Scientific American Books).

My own jointly edited book *The Cambridge Encyclopedia of Human Evolution*, ed. Steve Jones, Robert Martin and David Pilbeam (Cambridge University Press, 1992), has articles dealing with human and primate paleontology, comparative anatomy, anthropology and human genetics. A more technical account of many of these subjects is in *The Human Revolution*, ed. P. Mellors and C. Stringer (Edinburgh University Press, 1988). B. M. Fagan's *The Journey from Eden* (Thames and

Hudson, 1990) gives a romanticized but nevertheless fascinating account of humanity's spread over the world from the African homeland.

For engaging tales about evolution and the eccentrics who have studied it one cannot do better than *The Encyclopedia of Evolution: Humanity's Search for Its Origins* by Richard Milner (Facts on File, 1990). Almost everything in evolution descends from Darwin, and there is almost a Darwin industry in books about his life and work. The best recent biography and account of his work is *Darwin* by A. Desmond and J. Moore (Penguin Books, 1991). I have used language as a metaphor for evolutionary change throughout, and *The Cambridge Encyclopedia of Language* by David Crystal (Cambridge University Press, 1987) is a witty introduction to linguistics for those who, like me, come from outside the subject.

Anyone interested in what our own species has done to its environment as it spread across the world, and what it is likely to do in the future, should read *The Diversity of Life* by E. O. Wilson (Harvard University Press, 1993). Our evolutionary prospects are more likely to be determined by ecological disasters than by genetic change, and Wilson's book is a brilliant account of man's place in nature and of how we are in danger of ending our long genetic history by accident rather than by design.

For those who wish to pursue further the points raised in individual chapters, the following list may be of some help. It is far from exhaustive and, because of space constraints, many of the topics covered in the text are not referred to here. Wherever possible I have included the most recently published references consulted as these themselves lead back to the earlier work on the subject. I add the subject covered by a publication in parentheses when this is not obvious from its title.

INTRODUCTION THE FINGERPRINTS OF HISTORY

Keynes, M. 1993. *Sir Francis Galton, FRS. The Legacy of His Ideas*. Macmillan, London.

Kevles, D. 1986. *In the Name of Eugenics: Genetics and the Uses of Human Heredity*. Penguin, Harmondsworth.

Proctor, R. N. 1988. *Racial Hygiene: Medicine under the Nazis*. Harvard University Press.

Muller-Hill, B. 1988. *Murderous Science: Elimination by Scientific Selection of Jews, Gypsies and Others. Germany 1933–1945*. Oxford University Press.

CHAPTER 1   A MESSAGE FROM OUR ANCESTORS

Friar, S. 1987. *A New Dictionary of Heraldry.* Alpha Books, A and C Black, London.

Lewontin, R. 1982. *Human Diversity.* Scientific American Library, New York.

Spuhler, J. N. 1985. *Annual Review of Anthropology* 14: 103–33. "Anthropology, Evolution and Scientific Creationism."

Mollon, J. 1992. *Nature* 356: 378–79. "Worlds of Difference." (Inherited differences in color perception)

Bowman, J. E., and Murray, R. F. 1990. *Genetic Variation and Disorders in Peoples of African Origin.* Johns Hopkins University Press.

Nagel, R. L. et al. 1988. *New England Journal of Medicine* 312: 880–84. "Hemotologically and Genetically Distinct Forms of Sickle Cell Anemia in Africa."

Chakraborty, R. et al. 1992. *American Journal of Human Genetics* 50: 145–55. "Caucasian Genes in American Blacks: New Data."

Lawlor, D. A. et al. 1991. *Nature* 349: 785–87. "Ancient HLA Genes from 7,500-Year-Old Archaeological Remains."

Neel, J. V. et al. 1988. *American Journal of Human Genetics* 43: 870–93. "Protein Variation in Hiroshima and Nagasaki: Tales of Two Cities."

Cann, R. L. 1993. *Trends in Research on Ecology and Evolution* 8: 27–31. "Human Dispersal and Divergence." (Mitochondrial ancestry of human populations)

Sinclair, A. H. et al. 1990. *Nature* 346: 240–44. "A Gene from the Human Sex-Determining Region Encodes a Protein with Homology to a Conserved DNA-Binding Motif." (Gene for male sex-determination)

CHAPTER 2   THE RULES OF THE GAME

There are many introductory texts in genetics which deal with the basic rules of inheritance and of linkage. One of the most comprehensive and up-to-date is *An Introduction to Genetic Analysis* by A. J. F. Griffith and others (W. H. Freeman, New York, 5th edition, 1993). A. P. and E. J. Mange's *Basic Human Genetics* (Sinauer, Cambridge, Ma., 1993) gives an anthropocentric view of the subject. My own *Genetics for Beginners* (with Borin van Loon: Icon Books, London 1993), if nothing else, lives up to its title.

CHAPTER 3   HERODOTUS REVISED

Genes and Genomes by Maxine Singer and Paul Berg (University Science Books, Mill Valley, Ca., and Blackwell Scientific Publications, Oxford, 1991) is a solid treatment of modern genetics, much of which is derived from research on human genes. It deals with methods of mapping human genes. *Human Genetics: The Molecular Revolution* by Edwin H. McConkey (Jones and Bartlett, Baltimore, 1993) gives a clear account of how genes are mapped and discusses the relevance of this to medicine. Three articles by H. F. Judson, M. Gilbert and L. Hood in *The Code of Codes: Scientific and Social Issues in the Human Genome Project*, ed. D. J. Kevles and L. Hood (Harvard University Press 1993) discuss how physical mapping is carried out and what it is telling us.

The cystic fibrosis story is described in Tsui, L.-C. and Buchwald, M., 1991, *Advances in Human Genetics* 20: 153–266 "Biochemical and Molecular Genetics of Cystic Fibrosis." As gene mapping is proceeding so quickly it is difficult to keep up to date. The 1st October 1993 issue of *Science* is a "Genome Issue." The paper by Collins, F. and Galas, D. 1993 ("A New Five-Year Plan for the U.S. Human Genome Project." *Science* 262: 43–46) is particularly worth reading. Several of the papers in *The Code of Codes: Scientific and Social Issues in the Human Genome Project*, ed. D. J. Kevles and L. Hood (Harvard University Press, Cambridge, Ma., 1992) deal with mapping. That by Nancy Wexler ("Clairvoyance and Caution: Repercussions from the Human Genome Project") puts mapping into its social and scientific context.

CHAPTER 4   CHANGE OR DECAY

Montandon, A. J. et al. 1992. *Human Genetics* 89: 319–22. "Direct Estimate of the Haemophilia B Mutation Rate and Variation of the Sex-Specific Mutation Rates in Sweden."

Wallace, M. R. et al. 1991. *Nature* 353: 864–66. "A De Novo Alu Sequence Results in Neuro-Fibromatosis Type 1."

Harper, P. S. et al. 1992. *American Journal of Human Genetics* 51: 10–16. "Anticipation in Muscular Dystrophy; New Light on an Old Problem."

Sutherland G. R., and Richards, R. I. 1992. *American Journal of Human Genetics* 51: 7–9. "Anticipation Legitimized: Unstable DNA to the Rescue."

Neel, J. V. et al. 1990. *American Journal of Human Genetics* 46: 1053–72. "The Children of Parents Exposed to Atomic Bombs: Estimates of the Genetic Doubling Dose of Radiation for Humans."

Peto, J. 1990. *Nature* 345: 389–90. "Radon and the Risks of Cancer."

Clarke, R. H., and Southwood, T. R. E. 1989. *Nature* 338: 197–98. "Risks from Ionizing Radiation."

Evans, H. J. 1988. *Philosophical Transactions of the Royal Society of London* Series B 319: 325–40. "Mutation as a Cause of Genetic Disease."

Garner, C. 1992. *Nature* 360: 207–08. "Molecular Potential." (DNA adducts in Polish cities)

Hastie, N. D. et al. 1990. *Nature* 346: 865–68. "Telomere Reduction in Human Colorectal Cancer and with Ageing." (DNA loss with age)

CHAPTER 5   THE BATTLE OF THE SEXES

Maynard Smith, J. 1988. *Evolutionary Genetics*. Oxford University Press. (The origin and maintenance of sex)

Hurst, L. 1992. *Proceedings of the Royal Society of London* Series B 248: 135–48. "Intragenomic Conflict as an Evolutionary Force."

Dawkins, R. 1989. *The Selfish Gene*. Oxford University Press.

Dunbar, R. I. M. 1989. *Primate Social Systems*. Croom Helm, London.

Fisher, H. E. 1992. *Anatomy of Love: The Natural History of Monogamy, Adultery and Divorce*. W. W. Norton and Company, New York. (Comparative behavior of primates)

Johnson, A. M. et al. 1992. *Nature* 360: 410–13. "Sexual Lifestyles and HIV Risk." (Male and female mating frequencies)

Mulder, M. B. 1991. In *Behavioural Ecology*, ed. J. R. Krebs and N. B. Davies (Blackwell, Oxford), 69–98. "Human Behavioural Ecology." (Tribal peoples and male success in relation to wealth)

Spurdle, A., and Jenkins, T. 1992 *American Journal of Human Genetics* 50: 107–25. "Y Chromosome Probes Detect Complex Y Chromosome Haplotypes in Southern African Populations." (White males mating with black females)

Clarke, A. 1990. *Development* 1990 Supplement: 131–39. "Genetic Imprinting in Clinical Practice."

Moore T., and Haig, D. 1991. *Trends in Genetics* 7: 45–49. "Genomic Imprinting in Mammalian Development: A Parental Tug-of-War."

CHAPTER 6   CLOCKS, FOSSILS AND APES

Trinkaus, E., and Shipman, P. 1993. *The Neanderthals: Changing the Image of Mankind*. Jonathan Cape, London.

Brown, F. H. 1992. In *The Cambridge Encyclopedia of Human Evolution*, ed. Steve Jones, Robert Martin and David Pilbeam (Cambridge University Press). 179–87. "Dating Methods."

Simons, E. 1992. In *The Cambridge Encyclopedia of Human Evolution*, ed. Steve Jones, Robert Martin and David Pilbeam (Cambridge University Press). 199–209. "The Fossil History of Primates."

Lewin, R. 1993. *Human Evolution: An Illustrated Introduction* (3d edition). Blackwell Scientific Publications, Oxford.

Leakey, R., and Lewin, R. 1992. *Origins Reconsidered: In Search of What Makes Us Human*. Little, Brown & Co., New York.

Thorne, A. G., and Wolpoff, M. H. 1992. *Scientific American* 266: 28–33. "The Multiregional Evolution of Humans."

Caccone, A., and Powell, J. R. 1989. *Evolution* 43: 925–42. "DNA Divergence among Hominoids."

Friday, A. E. 1992. In *The Cambridge Encyclopedia of Human Evolution*, ed. Steve Jones, Robert Martin and David Pilbeam (Cambridge University Press). 295–99. "Measuring Relatedness: Calibrating the Molecular Clock."

Ou, C-Y et al. 1992. *Science* 256: 1165–71. "Molecular Epidemiology of HIV Transmission in a Dental Practice."

CHAPTER 7   TIME AND CHANCE

Lasker, G. W. 1989. *Surnames and Genetic Structure*. Cambridge University Press.

Bittles, A. H. et al. 1991. *Science* 252: 789–94. "Reproductive Behaviour and Health in Consanguineous Marriages."

Jorde, L. B., and Pitkanen, K. J. 1991. *American Journal of Physical Anthropology* 84: 127–50. "Inbreeding in Finland."

Williams, E. M. 1986. In *Genetic and Population Studies in Wales*, ed. P. S. Harper and E. Sunderland (University of Wales Press, Cardiff). 186–211. "Genetic Studies of Welsh Gypsies."

Stine, O. C., and Smith, K. D. 1990. *American Journal of Human Genetics* 46: 452–58. "The Estimation of Selection Coefficients in Afrikaners: Huntington Disease, Porphyria Variegata and Lipoind Proteinosis."

Jones, J. S., and Rouhani, S. 1986. *Nature* 319: 449–50. "Human Evolution: How Small Was the Bottleneck?"

Oakey, R., and Tyler-Smith, C. 1990. *Genomics* 7: 325–30. "Y-Chromosome Haplotyping Suggests that Most European and Asian Men are Descended from One of Two Males."

Hedges, S. B. et al. 1992. *Science* 255: 737–39. "Human Origins and Analysis of Mitochondrial DNA Sequences." (Criticisms of "mitochondrial Eve")

CHAPTER 8   THE ECONOMICS OF EDEN

Roberts, R. G. et al. 1990. *Nature* 345: 153–56. "Thermoluminescence Dating of a 50,000-year-old Human Occupation Site in Northern Australia."

Horai, S., and Hayasaka, K. 1990. *American Journal of Human Genetics* 46: 828–42. "Intraspecific Nucleotide Sequence Difference in the Major Non-Coding Regions of Human Mitochondrial DNA."

Vigilant, L. et al. 1991. *Science* 253: 1503–07. "African Populations and the Evolution of Human Mitochondrial DNA."

Hertzberg, M. et al. 1989. *American Journal of Human Genetics* 44: 504–10. "An Asian-Specific 9-bp Deletion of Mitochondrial DNA Is Frequently Found in Polynesians."

Hoffecker, J. F. et al. 1993. *Science* 259: 46–53. "The Colonization of Beringia and the Peopling of the New World."

Schurr, T. G. et al. 1990. *American Journal of Human Genetics* 46: 613–23. "Amerindian Mitochondrial DNAs Have Rare Asian Mutations at High Frequencies, Suggesting They Derived from Four Primary Maternal Lineages."

McCorriston, J., and Hole, F. 1991. *American Anthropologist* 93: 46–69. "The Ecology of Seasonal Stress and the Origin of Agriculture in the Near East."

Cohen, M. N. and Armelagos, G. J. 1991. *Paleopathology and the Origins of Agriculture.* Academic Press, London.

Zvelebil, M., and Dolukhanov, P. 1991. *Journal of World Prehistory* 5: 233–78. "The Transition to Farming in Eastern and Northern Europe."

Ammerman, A. J., and Cavalli-Sforza, L. L. 1984. *The Neolithic Transition and the Genetics of Populations in Europe.* Princeton University Press.

Bertranpetit, J., and Cavalli-Sforza, L. L. 1991. *Annals of Human Genetics* 55: 51–67. "A Genetic Reconstruction of the History of the Population of the Iberian Peninsula."

Antonarakis, S. E. et al. 1984. *Proceedings of the National Academy of Sciences* 81: 853–56. "Origin of the Beta-sickle Globin Gene in Blacks: The Contribution of Recurrent Mutation or Gene Conversion or Both."

CHAPTER 9   THE KINGDOMS OF CAIN

Park, T. 1992. *Journal of the American Anthropological Association* 94: 90–117. "Early Trends Towards Class Stratification: Chaos, Common Property and Flood Recession Agriculture."

Griffeth, R., and Thomas, C. G. 1981. *The City-State in Five Cultures.* ABC-Clio, Santa Barbara and London.

Piazza, A. et al. 1988. *Annals of Human Genetics* 52: 203–13. "A Genetic History of Italy."

Livshits, G., Sokal, R. R., and Kobyliansky, E. 1991. *American Journal of Human Genetics* 49: 131–46. "Genetic Affinities of Jewish Populations."

Li, H. et al. 1990. *Human Genetics* 86: 231–35. "Abnormal Hemoglobins in the Silk Road region of China."

Sokal, R. R. et al. 1990. *American Naturalist* 135: 157–75. "Genetics and Language in European Populations."

Cavalli-Sforza, L. L. 1991. *Scientific American* 265: 104–10. "Genes, People and Languages."

Gamkrelidze, T. V., and Ivanov, V. 1990. *Scientific American* 262: 82–89. "The Early History of Indo-European Languages."

Bellwood, P. 1991. *Scientific American* 265: 70–75. "The Austronesian Dispersal and the Origin of Languages."

Ross, P. E. 1991. *Scientific American* 264: 71–79. "Hard Words." (Reconstruction of early languages)

CHAPTER 10   DARWIN'S STRATEGIST

Fleischer, R. C., and Johnston, R. F. 1984. *Canadian Journal of Zoology* 62: 405–10. "The Relationships Between Winter Climate and Selection on Body Size in House Sparrows."

Foley, R. 1987. *Another Unique Species: Patterns in Human Evolutionary Ecology.* Longman; Harlow, Essex. (Climate, diet and human differentiation)

Guglielmino-Matessi, G. G. et al. 1979. *American Journal of Physical Anthropology* 50: 494–564. "Climate and the Evolution of Skull Metrics in Man."

Roberts, N. 1989. *The Holocene: An Environmental History.* Blackwell, Oxford.

Baker, H. T. 1992. In *The Cambridge Encyclopedia of Human Evolution,* ed. Steve Jones, Robert Martin and David Pilbeam (Cambridge University Press), pp. 46–51. "Human Adaptations to the Physical Environment."

Loomis, W. F. 1967. *Science* 157: 501–06. "Skin Pigment Regulation of Vitamin D Biosynthesis in Man."

Jones, J. S. 1982. *Nature* 298: 749–50. "Genetic Differences in Individual Behaviour Associated with Shell Polymorphism in the Snail *Cepaea nemoralis.*"

Johnson, R. C. et al. 1987. *Human Biology* 53: 1–14. "Genetic Interpretation of Racial/Ethnic Differences in Lactose Absorption: A Review."

CHAPTER 11   THE DEADLY FEVERS

Crosby, A. W. 1993. *Ecological Imperialism: The Biological Expansion of Europe, 900–1900.* Cambridge University Press/Canto Books.

McKeown, T. 1988. *The Origins of Human Disease.* Basil Blackwell, Oxford.

Kiple, K. F. (ed.) 1993. *The Cambridge World History of Disease.* Cambridge University Press.

Anderson, R. M., and May, R. M. 1991. *Infectious Diseases of Humans: Dynamics and Control.* Oxford University Press.

Barbour, A. G., and Fish, D. 1993. *Science* 260: 1610–17. "The Biological and Social Phenomenon of Lyme Disease."

Krause, R. M. 1992. *Science* 257: 1073–78. "The Origin of Plagues, Old and New."

Waters, A. P. et al. 1991. *Proceedings of the National Academy of Sciences* 88: 3140–44. "*Plasmodium falciparum* Appears to Have Arisen as a Result of Lateral Transfer Between Avian and Human Hosts."

Desowitz, R. S. 1993. *The Malaria Capers: More Tales about Parasites and People.* W. W. Norton and Company, New York.

Hill, A. V. S. et al. 1992. *Proceedings of the National Academy of Sciences* 89: 2277–

81. "Extensive Genetic Diversity in the HLA Class II Region of Africans, with a Focally Predominant Allele DRB*1304." (Genetics of resistance to malaria)

CHAPTER 12   CALIBAN'S REVENGE

Grandjean, P. 1991. *Ecogenetics: Genetic Predisposition to the Toxic Effects of Chemicals*. Chapman and Hall, London.

Varmus, H. E., and Weinberg, R. A. 1993. *Genes and the Biology of Cancer*. Scientific American Library, W. H. Freeman and Co., New York.

Goddard, A. D., and Solomon, E. 1992. *Advances in Human Genetics* 21: 321–58. "Genetic Aspects of Cancer."

McCluer, J. W., and Kammerer, C. M. 1991. *American Journal of Human Genetics* 49: 1139–44. "Dissecting the Genetic Component to Coronary Heart Disease."

Stinson, S. 1992. *Annual Review of Anthropology* 21: 143–70. "Nutritional Adaptation." (Diet, genetics and health)

Long, J. C. et al. 1991. *American Journal of Physical Anthropology* 84: 141–57. "Genetic Variation in Arizona Mexican Americans: Estimation and Interpretation of Admixture Proportions." (Diabetes and Amerindian ancestry)

Bouchard, T. J. et al. 1990. *Science* 250: 223–28. "Sources of Human Psychological Differences: The Minnesota Study of Twins Reared Apart."

Keller, E. F. 1992. In *The Code of Codes: Scientific and Social Issues in the Human Genome Project*, ed. D. J. Kevles and L. Hood (Harvard University Press). 281–99. "Nature, Nurture and the Human Genome Project."

Hamer, D. H. et al. 1993. *Science* 261: 321–25. "A Linkage Between DNA Markers on the X Chromosome and Male Sexual Orientation."

Brunner, H. G. et al. 1993. *Science* 262: 578–80. "Abnormal Behavior Associated with a Point Mutation in the Structural Gene for Monoamine Oxidase A."

CHAPTER 13   COUSINS UNDER THE SKIN

Baker, J. R. 1974. *Race*. Oxford University Press.

Banton, M. 1987. *Racial Theories*. Cambridge University Press.

Reynolds, V., Falger, V. S. E., and Vine, I. 1987. *The Sociobiology of Ethnocentrism*. Croom Helm, London.

Ludmerer, K. 1992. *Genetics and American Society: A Historical Appraisal*. Johns Hopkins University Press, Baltimore.

Thomassen, H. R. et al. 1991. *American Journal of Human Genetics* 48: 677–81. "Alcohol and Aldehyde Dehydrogenase Genotypes and Alcoholism in Chinese men."

Excoffier, L. et al. 1987. *Yearbook of Physical Anthropology* 30: 151–94. "Genetics and History of Sub-Saharan Africa."

Latter, B. D. H. 1980. *American Naturalist* 116: 220–37. "Genetic Differences Within and Between Populations of the Major Human Subgroups."

Nei, M., and Roychoudhury, A. K. 1982. *Evolutionary Biology* 14: 1–59. "Genetic Relationship and Evolution of Human Races."

Devlin, B., and Risch, N. 1992. *American Journal of Human Genetics* 51: 534–48. "Ethnic Differentiation at VNTR Loci, with Special Reference to Forensic Applications." (DNA fingerprints in human races)

Balazs, I. et al. 1992. *Genetics* 131: 191–98. "Human Population Genetic Studies Using Hypervariable Loci. I. Analysis of Assamese, Australian, Cambodian, Caucasian, Chinese and Melanesian Populations." (Global differentiation of DNA fingerprints)

CHAPTER 14   EVOLUTION ENGINEERED

Tudge, C. 1993. *The Engineer in the Garden: From the Idea of Heredity to the Creation of Life.* Jonathan Cape, London.

Bud, R. 1992. *The Uses of Life: A History of Biotechnology.* Cambridge University Press.

Bains, W. 1993. *Biotechnology from A to Z.* Oxford University Press.

Fowler, C., and Mooney, P. 1990. *The Threatened Gene: Food, Politics and the Loss of Genetic Diversity.* Lutterworth Press, Cambridge.

CHAPTER 15   FEAR OF FRANKENSTEIN

Chadwick, R. 1987. *Ethics, Reproduction and Genetic Control.* Routledge, London and New York.

Clarke, A. 1991. *Lancet* 338: 998–1000. "Is Non-Directive Genetic Counselling Possible?"

Caskey, C. T. 1993. *Science* 262: 48–57. "Presymptomatic Diagnosis: A First Step Toward Genetic Health Care."

Marteau, T. M. et al. 1992. *Journal of Medical Genetics* 29: 24–26. "Effects of Genetic Screening on Perceptions of Health: A Pilot Study." (Sickle Cell and screening)

McNeil, T. F. et al. 1988. *Thorax* 43: 505–07. "Psychosocial Effects of Screening for Somatic Risk: The Swedish Alpha-1-Antitrypsin Experience."

Watson, E. K. et al. 1991. *British Medical Journal* 303: 504–07. "Screening for Carriers of Cystic Fibrosis Through Primary Health Care Services."

Tyler, A. et al. 1992. *British Medical Journal* 304: 1593–96. "Pre-Symptomatic Testing for Huntington's Disease in the United Kingdom."

Anderson, W. F. 1992. *Science* 256: 808–13. "Human Gene Therapy."

Mulligan, R. C. 1993. *Science* 260: 926–32. "The Basic Science of Gene Therapy."

Wivel, N. A., and Walters, L. 1993. *Science* 262: 533–38. "Germ-Line Gene Modification and Disease Prevention: Some Medical and Ethical Perspectives."

Ostner, H. et al. 1993. *American Journal of Human Genetics* 523: 565–77. "Insurance and Genetic Testing: Where Are We Now?"

CHAPTER 16   THE EVOLUTION OF UTOPIA

Vogel, F. 1992. *Human Genetics* 89: 127–46. "Risk Calculations for Hereditary Effects of Ionizing Radiation in Humans."

Modell, B., and Kuliev, A. M. 1989. *Clinical Genetics* 36: 286–98. "Impact of Public Health on Human Genetics." (Age and the rate of mutation; inbreeding and outbreeding)

Ulizzi, L., and Terrenato, L. 1992. *Annals of Human Genetics* 56: 113–18. "Natural Selection Associated with Birth Weight. VI. Towards the End of the Stabilising Component."

Reddy, B. M., and Chopra, V. P. 1990. *American Journal of Physical Anthropology* 83: 281–96. "Opportunity for Natural Selection among the Indian Populations."

Fries, J. F. et al. 1989. *Lancet* 1989 (1): 481–83. "Health Promotion and the Compression of Morbidity."

Olshansky, S. J. et al. *Science* 250: 634–40. "In Search of Methuselah: Estimating the Upper Limits to Human Longevity."

Ellison, P. T. 1991. In *Applications of Biological Anthropology to Human Affairs*, ed. C. G. N. Mascie-Taylor and G. W. Lasker (Cambridge University Press). 14–54. "Reproductive Ecology and Human Fertility."

Chakraborty, R. 1986. *Yearbook of Physical Anthropology* 29: 1–43. "Gene Admixture in Human Populations: Models and Predictions."

Bittles, A. H. et al. 1991. *Science* 252: 789–94. "Reproductive Behavior and Health in Consanguineous Marriages."

Caldwell, J. C., and Caldwell, P. 1990. *Scientific American* 262: 82–89. "High Fertility in Sub-Saharan Africa."

Index